H.L. Mencken

A
Religious
Orgy
in
Tennessee

A
Religious
Orgy
in
Tennessee

A Reporter's Account of
the Scopes Monkey Trial

H.L. Mencken

Introduction by Art Winslow

MELVILLE HOUSE PUBLISHING
BROOKLYN, NEW YORK

Articles originally published in *The Baltimore Sun* are © 1925 and reprinted with the permission of the *The Baltimore Sun*. Permission also granted by Enoch Pratt Free Library, Baltimore, in accordance with the terms of the bequest of H.L. Mencken. "In Tennessee" is reprinted from the July 1, 1925, issue of *The Nation*. Permission also granted by Enoch Pratt Free Library, Baltimore, in accordance with the terms of the bequest of H.L. Mencken. Permission to reprint "To Expose a Fool" from *The American Mercury*, October 1925, is granted by Enoch Pratt Free Library, Baltimore, in accordance with the terms of the bequest of H.L. Mencken.

Cover photo shows H.L. Mencken (upper right with hands clasped behind his back) standing on a table to look over the crowd at the proceedings at the Rhea County Courthouse. This photograph is courtesy of the Enoch Pratt Free Library, H.L. Mencken Collection.

Special acknowledgement is made to Averil Kadis, of the Enoch Pratt Free Library, and Tom Davis, of Bryan College, for assistance with research and permissions.

Book Design: David Konopka

Melville House Publishing
145 Plymouth Street
Brooklyn, New York 11201

Printed in the United States of America.

Library of Congress Cataloging-in-Publication Data

Mencken, H. L. (Henry Louis), 1880-1956.
  "A religious orgy in Tennessee" : a reporter's account of the Scopes "monkey" trial / H.L. Mencken. — 1st ed.
      p. cm.
  Articles originally published in The Baltimore Sun, The Nation, or The American Mercury.
  ISBN-13: 978-1-933633-17-6 (alk. paper)
  ISBN-10: 1-933633-17-4 (alk. paper)
  1. Scopes, John Thomas—Trials, litigation, etc.—Press coverage—Maryland—Baltimore. 2. Evolution—Study and teaching—Law and legislation—Tennessee. 3. Mencken, H. L. (Henry Louis), 1880-1956—Political and social views. I. Title.
KF224.S3M46 2006
345.73'0288—dc22
                                        2006022870

Introduction *by Art Winslow*   ix

I: The Tennessee Circus   3

II: Homo Neanderthalensis   11

III: In Tennessee   19

IV: Mencken Finds Daytonians Full of   27
        Sickening Doubts About Value of Publicity

V: Impossibility of Obtaining Fair Jury   35
        Insures Scopes' Conviction, Says Mencken

VI: Mencken Likens Trial to a Religious Orgy,   41
        with Defendant a Beelzebub

VII: Yearning Mountaineers' Souls Need   49
        Reconversion Nightly, Mencken Finds

VIII: Darrow's Eloquent Appeal Wasted on Ears   61
        That Heed Only Bryan, Says Mencken

IX: Law and Freedom, Mencken Discovers,   67
        Yield Place to Holy Writ in Rhea County

X: Mencken Declares Strictly Fair Trial Is   75
        Beyond Ken of Tennessee Fundamentalists

XI: Malone the Victor, Even Though Court   81
        Sides with Opponents, Says Mencken

XII: Battle Now Over, Mencken Sees;   89
        Genesis Triumphant and Ready for New Jousts

XIII: Tennessee in the Frying Pan   95

XIV: Bryan   103

XV: Round Two   111

XVI: Aftermath   119

XVII: To Expose a Fool   127

        Photographs   137

        Appendix: The Examination of   147
        William Jennings Bryan by Clarence Darrow

Introduction

By Art Winslow

We have the stage antics of baseball-player-turned-evangelist William Ashley Sunday—Billy Sunday—and the weak ankles of the Tennessee General Assembly to thank for the high theater of the Scopes Trial of 1925. As recounted by Edward J. Larson in his excellent *Evolution: The Remarkable History of a Scientific Theory*, the state senate, considering a ban on the teaching of evolution in public schools, had already rejected the idea in committee when Billy Sunday swept into Memphis for an eighteen-day revival that February.

Long before Elvis was to earn fame with his gyrations, this gymnast for Jesus, Larson tells us, "jumped, kicked and slid across the stage" while denouncing the "tommyrot" of evolution, the possibility "that we came from protoplasm, instead of being born of God Almighty." When the legislators noticed that Sunday's sermons had drawn an aggregate audience of some 200,000 constituents, the senate committee reversed itself, leading to passage of the Butler Act, signed into law in March by Governor Austin Peay virtually unchanged from the wording crafted by its author, a farmer named John Washington Butler.

"It shall be unlawful for any teacher in any of the universities, Normals and all other public schools of the State which are supported in whole or in part by the public school funds of the State, to teach any theory that denies the story of the Divine Creation of man as taught in the Bible, and to teach instead that man has descended from a lower order of animals," the Butler Act stipulated. To transgress it was a misdemeanor, punishable by a minimum fine of $100 and a maximum of $500 for each offense.

Tennessee was not the only state in which sentiment against Darwin's ideas soaked enough ground to seep into public policy. Two years before, in 1923, Oklahoma had banned textbooks that promoted Darwinism, and in 1924 California required teachers to approach evolution as a theory only. Kansas was a seedbed for anti-evolution activists as well, including the lecturer Charles L. Clayton, who tried to blur the boundaries between science and religion by asserting an intelligent force, a "Doctrine of Design," which can be seen as the forebear of modern-day belief in "Intelligent Design."

One of the preeminent leaders of the anti-Darwin movement at the time was William Jennings Bryan, a three-time presidential candidate, two-time Nebraska Congressman and onetime Secretary of State for Woodrow Wilson. Known as the "Great Commoner," Bryan's 1922 book *In His Image* made clear his feelings on the subject of evolution:

> While "survival of the fittest" may seem plausible
> when applied to individuals of the same species,
> it affords no explanation whatever, of the almost
> infinite number of creatures that have come under
> man's observation. To believe that natural selection,
> sexual selection or any other kind of selection can
> account for the countless differences we see about
> us requires more faith in chance that a Christian
> is required to have in God.

*   *   *

The Bible does not say that reproduction shall be nearly according to kind or seemingly according to kind. The statement is positive that it is according to kind, and that does not leave room for the changes however gradual or imperceptible that are necessary to support the evolutionary hypothesis.

Bryan also wrote, "If we accept the Bible as true we have no difficulty in determining the origin of man," and that Darwinian doctrine "has been the means of shaking the faith of millions," is "absurd and harmful to society," and further, it "attacks the very foundations of Christianity." Darwinism "offers no reason for existence and presents no philosophy of life; the Bible explains why man is here and gives us a code of morals that fits into every human need," Bryan observed.

Bryan's involvement in the Scopes trial had an air of preordainment about it, for the proceeding itself smacked of intentionality from the outset. In early May 1925, the fledgling American Civil Liberties Union publicized its eagerness to find a test case against the newly minted Butler Act by offering to defend any teacher accused under it. George Rappelyea, a manager for the Cumberland Coal and Iron company in Dayton, Tennessee, saw notice of that in the *Chattanooga Daily Times* and, sensing opportunity, brought it to the attention of Frank Earle Robinson, proprietor of a local drug store

but also chair of the Rhea County School Board. Before
another day had elapsed, a meeting was arranged that
included Rappelyea, Robinson, school superintendent
Walter White, a pair of the city's lawyers (Sue Hicks and
his brother Herbert Hicks, who went on to work for the
prosecution), and John Scopes, an athletic coach and
substitute teacher who agreed to be the defendant to test
the law. Scopes was not the school's designated biology
teacher, but he had filled in and perhaps even taught
evolution in a way that contravened the law (the defense
did not contest the matter at trial).

Generation of publicity was the primary motivation,
and that came with a vengeance. A warrant was sworn
out, the A.C.L.U. was notified, and so were members of
the press. Various parties, including the Christian
Fundamentals Association, quickly urged Bryan to join
the cause, and in a week he did, saying, "They came here
to try revealed religion. I am here to defend it." Bryan
aided lead prosecutor A.T. Stewart, who was attorney
general for the Eighteenth Judicial Circuit and a future
Senator. Rounding out the team was Bryan's son, the
Hicks siblings, and a former Dayton assistant attorney
general, Ben McKenzie, and his son as well.

On the opposing side, Clarence Darrow, perhaps the
country's best-known trial lawyer, agreed to join the
A.C.L.U.'s Arthur Garfield Hays and home-state eminence

John R. Neal, a law school dean from Knoxville, in defending Scopes. ("Scopes isn't on trial. Civilization is on trial," Darrow quipped.) Also brought on board was Darrow friend and prominent divorce lawyer Dudley Field Malone—who had served as Assistant Secretary of State under Bryan in the Wilson administration. Overseeing the trial was the presiding judge of the Eighteenth Circuit, John Raulston, a devout Baptist.

And then there was H.L.Mencken. As an unattributed witticism in Marion Rodgers's *Mencken: The American Iconoclast* put it, "If the Scopes Trial had not existed, H.L. Mencken would have had to invent it." Scopes himself went so far as to suggest, four decades after the trial, that "In a way it was Mencken's show," the journalist's lacerating critique of the Bible Belt (a term Mencken coined, along with "booboisie") garnering most of the lasting attention.

Mencken's exact relation to the defense team in the Scopes trial remains somewhat vague, but it certainly transcended boundaries that would be considered proper for a journalist today. Rodgers reports that Hays asked Mencken to help with the Scopes defense strategy. Mencken's other recent biographer, Terry Teachout (*The Skeptic*), writes that "there is no question that he provided pre-trial advice to the A.C.L.U.," but Mencken's claim that it was his idea to put Bryan on the stand,

and the suggestion that he helped persuade Darrow to join the defense team, may be overstatements. By all accounts, Mencken considered the ruination of Bryan to be a goal that eclipsed the protection of a schoolteacher, fitting for someone who considered his journalistic aim "to combat, chiefly by ridicule, American piety, stupidity, tin-pot morality, cheap chauvinism."

Mencken was the ultimate crusader, estimating late in his career that he had written more than 5 million words in his nineteen books and as a reporter, critic, columnist and editorialist for Baltimore's "Sunpapers" and the magazines *Smart Set* and the *American Mercury*, the last of which he not only edited but co-founded. Much of his prodigious energy was spent critiquing religion, particularly fundamentalist Protestantism, and its effect on civic life; he felt his own messianic calling in battling ignorance whenever he saw it infecting the public sphere. He was a believer in higher and lower orders of humanity, and penned columns that can be especially startling to encounter today, laced with racist and anti-Semitic or class-based bias.

The great paradox of Mencken is irreducible. The thrilling energy of his prose, the lucidity of his thought, the striking humor, the unabashed honesty and lack of restraint, the fearlessness to say what he thought no matter the offense that might be taken—those created a deeply admiring readership. The biliousness of his attacks, the withering scorn he subjected so many to in various group libels—those created a sense of hatred. Teachout reports that in 1926, a year after the Scopes trial, roughly 500 editorials about Mencken were published in American newspapers, and four-fifths of those were unfavorable.

To read Mencken is to understand why columnist Walter Lippmann would remark, in the *Saturday Review of Literature,* that "He calls you a swine, and an imbecile, and he increases your will to live," and yet, "What Mr. Mencken has created is a personal force in American life which has an extraordinarily cleansing and vitalizing effect." The humorist S.J. Perelman called Mencken "the ultimate firework" and credited him with loosening up journalism from its gray moorings. When Mencken's diary was published in 1989 and raised the ire of many, he was defended in a group letter to the *New York Review of Books* signed by Ralph Ellison, Norman Mailer, Arthur Miller, William Styron, Kurt Vonnegut and others, who hailed him as "a tremendous liberating force in American culture."

In his essays and reporting, Mencken comes off as an unusual fusion of anti-authoritarian and demotic impulses, disdainful of the crowd and group-think. In a piece titled "The Foundations of Quackery," which appeared in the *Baltimore Evening Sun* two years before the Scopes trial, he remarks "I have often pointed out how politics, under democracy, invariably translates itself from the domain of logical ideas to the domain of mere feelings, usually simple fear." That conviction—that it was simple fear being played upon, in this instance by the anti-Darwin movement—drove much of his coverage of the Scopes trial, just as it inflected his writing lifelong.

The Scopes trial lasted eight days in courtroom time, eleven days by the calendar, July 10 – 21 in stifling heat. Mencken's reportage covers many of the oddities of the setting and proceedings: the carnival atmosphere that surrounded the courthouse, the tussle over prayers at the opening of each court day, Judge Raulston's decision to disallow expert testimony before the jury (he allowed some to be read into the record). That may have been the stroke that caused Mencken to abandon Dayton on July 18, before the trial ended. In his memoirs, Mencken reported leaving to avoid dereliction of his duties to the *American Mercury*, but his departure meant that he missed Darrow's examination of Bryan on the stand, called there by the defense to serve as an expert witness on the Bible. The parrying between Darrow and Bryan in the two-hour exchange appears in the appendix of this book, the trial transcript perhaps most remarkable for what Bryan tries mightily to avoid saying.

In the end, despite the publicity it generated, what the 3,000 curious outsiders who flocked to Dayton saw was a trial whose cultural repercussions far exceeded its legal import. Rather than managing to hook the large constitutional issues it could have—concerning the validity of the restriction on teaching evolution—it was limited to a finding of whether or not John Scopes had violated the Butler Act. He had, even though the textbook he used, George Hunter's *Civic Biology,* had been approved by the state; it took the jury only nine minutes to decide on a guilty verdict. Scopes was fined $100. On appeal, in 1927 the Tennessee Supreme Court overturned it on a technicality: the judge had determined the penalty rather than the jury, which had the right under the state constitution. Since Scopes was no longer employed by the State of Tennessee, the court remarked, "We see nothing to be gained by prolonging the life of this bizarre case" and recommended to Attorney General Stewart that he drop the matter, which he did, ending Darrow's hopes that the case could be brought to the Supreme Court.

William Jennings Bryan died five days after the end of the trial. The full fury of what Mencken felt is represented in the original obituary essay he wrote, which appeared in the *Baltimore Evening Sun* on July 27 and is reprinted here. The critic Alfred Kazin, in his *On Native Grounds,* said that while Mencken's essay on Bryan was "one of the cruelest things he ever wrote," it "was probably the most brilliant" as well. For one of the few times in his life, though, Mencken censored himself due to an immediate outcry, and produced a softer, second version for the paper. His rethinking appeared in the *American Mercury* in October, and can be found in the ending of these pages. Even in giving his own prose a purgative scrub, Mencken retains his punch: "If the fellow was sincere, then so was P.T. Barnum," he wrote of Bryan.

The unresolved issues at the heart of the Scopes trial remained vexing questions for decades. Mississippi and Arkansas went on to pass statutes inspired by the Butler Act, and other states imposed less onerous restrictions on the teaching of evolution.

In *Epperson v Arkansas* in 1968, however, the Supreme Court overruled the Arkansas Supreme court, finding that the language so closely modeled on Butler's (making it unlawful in a state-supported school "to teach the theory or doctrine that mankind ascended or descended from a lower order of animals," or to adopt or use a textbook that did) "must be stricken because of its conflict with the constitutional prohibition of state laws respecting an establishment of religion or prohibiting the free exercise thereof." (Susan Epperson was a high school teacher from Little Rock, contesting the law.) Justice Abe Fortas, who wrote the majority opinion, stated "there can be no doubt that Arkansas has sought to prevent its teachers from discussing the theory of evolution because it is contrary to the belief of some that the Book of Genesis must be the exclusive source of doctrine as to the origin of man.... Plainly, the law is contrary to the mandate of the First, and in violation of the Fourteenth, Amendment to the Constitution."

Almost two decades later, in 1987, the Supreme Court invalidated a Creationism Act that had been passed in the state of Louisiana. In *Edwards v Aguillard*, the parents of public school children challenged the statute, the state represented by Governor Edwin Edwards. The state law required that if evolution was taught, it must be accompanied by instruction in "creation science." The act was found in violation of the Establishment Clause of the First Amendment. Justice William Brennan, writing for the majority, affirmed an appellate court conclusion that "the Act does not serve to protect academic freedom, but has the distinctly different purpose of discrediting evolution by counterbalancing its teaching at every turn with the teaching of creationism." Further, he noted, "the preeminent purpose of the Louisiana Legislature was clearly to advance the religious viewpoint that a supernatural being created humankind."

Such legal remedies have interrupted but hardly stopped the flow of news stories that snap us back to visions of the Scopes trial. In August of 1999, the Kansas Board of Education, ignoring the advice of a twenty-seven member advisory board of scientists, voted to adopt new science standards and leave it to local school boards whether or not to teach general Darwinian views, and further to strike knowledge of macroevolution from state assessment tests. Teaching the "Big Bang" theory and reference to the age of the earth were also left to local discretion. Most of the anti-evolution members of the state board were voted out in 2000, and in 2001 the new board reinstated the evolution requirement in the state's science curriculum.

In Pennsylvania, a local school board voted in 2004 to institute a requirement for teachers to read a statement about "intelligent design" prior to teaching evolution in high school biology, which led several parents to file suit. That suit, *Kitzmiller v Dover Area School District*, was decided in December 2005, with U.S. District Judge John Jones finding the board's actions unconstitutional under the Establishment Clause of the First Amendment, and further determining that "intelligent design" is creationism and not science. (School board members who were involved in the decision were also voted out of office, subsequently.)

In 2005, Bruce Alberts, president of the National Academy of Sciences, went so far as to post a warning about "increasing challenges to the teaching of evolution in public schools," and the inclusion of non-scientifically based "alternatives" in science courses. He noted that educators, according to news reports, were "quietly being urged to avoid teaching about evolution," even where controversy was not overt. Alberts issued an appeal to those outside the life sciences, fearing that the trend will spread to the earth and physical sciences as well. Mencken, it should be noted, wrote after Bryan's death that "the fire is still burning on many a far-flung hill, and it may begin to roar again at any moment."

"Democracy, in the last analysis, is only a sort of dream," Mencken wrote in his notebooks (collected in the book *Minority Report*). "It should be put in the same category as Arcadia, Santa Claus and Heaven. It is always a mistake to think of it as a reality. It never really exists; it is simply a forlorn hope." Lucky for us, he kept hoping.

A
Religious
Orgy
in
Tennessee

I

The
Tennessee
Circus

From *The Baltimore Evening Sun*, June 15, 1925

I

It is an old and bitter observation that, in armed conflicts, the peacemaker frequently gets the worst of it. The truth of the fact is being demonstrated anew in the case of the Tennessee pedagogue accused of teaching Evolution. No matter what the issue of that great moral cause, it seems to me very unlikely that either of the principal parties will be greatly shaken. The Evolutionists will go on demonstrating, believing in and teaching the mutability of living forms, and the Ku Klux theologians will continue to whoop for Genesis undefiled. But I look for many casualties and much suffering among the optimistic neutrals who strive to compose the controversy—that is, among the gentlemen who believe fondly that modern science and the ancient Hebrew demonology can be reconciled.

This reconciliation will take place, perhaps, on that bright day when Dr. Nicholas Murray Butler and the Hon. Wayne B. Wheeler meet in a saloon under a Baptist church, and drink *Brüderschaft* in a mixture of Clos Vougeot and Coca-Cola. But not before. For the two parties, it must be manifest, are at the farthermost poles of difference, and leaning out into space. If one of them is right at all, then the other is wrong altogether. There can be no honest compromise between them. Either Genesis embodies a mathematically accurate statement of what took place during the week of June 3, 4004 BC or Genesis is not actually the Word of God. If the former alternative be accepted, then all of modern science is nonsense; if the latter, then evangelical Christianity is nonsense.

This fact must be apparent, I believe, to everyone who has given sober and prayerful thought to the controversy. It should be especially apparent to those who now try to talk it away. I have, I confess, a great suspicion of such persons. When they pretend to be scientists it always turns out on inspection that they are only half-scientists—that no fact, however massive, is yet massive enough to keep them off the mourners' bench. And when they pretend to be Christians they are always full of mental reservations, which is to say, they are full of secret doubts, heresies and hypocrisies.

II

When I say Christians, of course, I mean Christians of the sort who accept the Bible as their sole guide to the divine mysteries, and are forced, in consequence, to take it exactly as it stands. There are also, of course, persons of the name who subscribed to arier and more sophisticated cults, each with its scheme for ameliorating the disconcerting improbability of certain parts of Holy Writ. Some of these cults get around the difficulty by denying that any sort of belief whatever, save perhaps in a few obvious fundamentals, is necessary to the Christian way of life—that a Christian is properly judged not by what he believes, but by what he does. And others dispose of the matter by setting up an authority competent to "interpret" the Scriptures, *i.e.*, to determine, officially and finally, what they mean or ought to mean when what they say is obscure or incredible.

Of the latter cults the most familiar is the Roman Catholic. It does not reject or neglect the Bible, as the Ku Klux Protestants allege; it simply accepts frankly the obvious fact that the Bible is full of difficulties—or, as the non-believer would say, contradictions and absurdities. To resolve these difficulties it maintains a corps of experts specially gifted and trained, and to their decision, when reached in due form of canon law, it gives a high authority. The first of such experts, in normal times, is the Pope; when he settles a point of doctrine, *i.e.*, of Biblical interpretation, the faithful are bound to give it full credit. If he is in doubt, then he may summon a Council of the Church, *i.e.*, a parliament of all the chief living professors of the divine intent and meaning, and submit the matter to it. Technically, I believe, this council can only advise him; in practice, he usually follows the view of its majority.

The Anglican, Orthodox Greek and various other churches, including the Presbyterian, follow much the same plan, though with important differences in detail. Its defects are not hard to see. It tends to exalt ecclesiastical authority and to discourage the study of Holy Writ by laymen. But its advantages are just as apparent. For one thing, it puts down amateur theologians, and stills their idiotic controversies. For another thing, it quietly shelves the highly embarrassing questions of the complete and literal accuracy of the Bible. What has not been singled out for necessary belief, and interpreted by authority, is tacitly regarded as not important.

III

Out of this plan flows the fact that the Catholics and their allies, in the present storm, are making much better weather of it than the evangelical sects. Their advantage lies in the simple fact that they do not have to decide either for Evolution or against it. Authority has not spoken upon the subject; hence it puts no burden upon conscience, and may be discussed realistically and without prejudice. A certain wariness, of course, is necessary. I say that authority has not spoken; it may, however, speak to-morrow, and so the prudent man remembers his step. But in the meanwhile there is nothing to prevent him examining all the available facts, and even offering arguments in support of them or against them—so long as those arguments are not presented as dogma.

The result of all this is that the current discussion of the Tennessee buffoonery, in the Catholic and other authoritarian press, is immensely more free and intelligent than it is in the evangelical Protestant press. In such journals as the *Conservator*, the new Catholic weekly, both sides are set forth, and the varying contentions are subject to frank and untrammeled criticism. Canon de Dorlodot whoops for Evolution; Dr. O'Toole denounces it as nonsense. If the question were the Virgin Birth, or the apostolic succession, or transubstantiation, or even birth control, the two antagonists would be in the same trench, for authority binds them there. Bill on Evolution authority is silent, and so they have at each other in the immemorial manner of theologians, with a great kicking up of dust.

The *Conservator* itself takes no sides, but argues that Evolution ought to be taught in the schools—not as incontrovertible fact but as a hypothesis accepted by the overwhelming majority of enlightened men. The objections to it, theological and evidential, should be noted, but not represented as unanswerable.

IV

Obviously, this is an intelligent attitude. Equally obviously, it is one that the evangelical brethren cannot take without making their position absurd. For weal or for woe, they are committed absolutely to the literal accuracy of the Bible; they base their whole theology upon it. Once they admit, even by inference, that there may be a single error in Genesis, they open the way to an almost complete destruction of that theology. So they are forced to take up the present challenge boldly, and to prepare for a battle to the death. If, when and as they attempt a compromise, they admit defeat.

Thus there is nothing unnatural in their effort to protect their position by extra-theological means—for example, by calling in the law to put down their opponents. All Christians, when one of their essential dogmas seems to be menaced, turn instinctively to the same device.

The whole history of the church, as everyone knows, is a history of schemes to put down heresy by force. Unluckily, those schemes do not work as well as they did in former ages. The heretic, in the course of time, has learned how to protect himself—even how to take the offensive. He refuses to go docilely to the stake. Instead, he yells, struggles, makes a frightful pother, bites his executioner. The church begins to learn that it is usually safest to let him go.

The Ku Klux Klergy, unfortunately for their cause, have not yet mastered that plain fact. Intellectually, there are still medieval. They believe that the devices which worked in the year 1300 will still work in 1925. As a life-long opponent of their pretensions I can only report that their fidelity to this belief fills me with agreeable sentiments. I rejoice that they have forced the fighting, and plan to do it in the open. My prediction is that, when the peanut shells are swept up at last and the hot-dog men go home, millions of honest minds in this great republic, hirtheto uncontaminated by the slightest doubt, will have learned to regard parts of Genesis as they now regard the history of Andrew Gump.*

* Andrew Gump was a lead character in a long running family comic strip.

II

Homo
Neanderthalensis

From *The Baltimore Evening Sun*, June 29, 1925

I

Such obscenities as the forthcoming trial of the Tennessee evolutionist, if they serve no other purpose, at least call attention dramatically to the fact that enlightenment, among mankind, is very narrowly dispersed. It is common to assume that human progress affects everyone—that even the dullest man, in these bright days, knows more than any man of, say, the Eighteenth Century, and is far more civilized. This assumption is quite erroneous. The men of the educated minority, no doubt, know more than their predecessors, and of some of them, perhaps, it may be said that they are more civilized—though I should not like to be put to giving names—but the great masses of men, even in this inspired republic, are precisely where the mob was at the dawn of history. They are ignorant, they are dishonest, they are cowardly, they are ignoble. They know little if anything that is worth knowing, and there is not the slightest sign of a natural desire among them to increase their knowledge.

Such immortal vermin, true enough, get their share of the fruits of human progress, and so they may be said, in a way, to have their part in it. The most ignorant man, when he is ill, may enjoy whatever boons and usufructs modern medicine may offer—that is, provided he is too poor to choose his own doctor. He is free, if he wants to, to take a bath. The literature of the world is at his disposal in public libraries. He may look at works of art. He may hear good music. He has at hand a thousand devices for making life less wearisome and more tolerable: the telephone, railroads, bichloride tablets, newspapers, sewers, correspondence schools, delicatessen. But he had no more to do with bringing these things into the world than the horned cattle in the fields, and he does no more to increase them today than the birds of the air.

On the contrary, he is generally against them, and sometimes with immense violence. Every step in human progress, from the first feeble stirrings in the abyss of time, has been opposed by the great majority of men. Every valuable thing that has been added to the store of man's possessions has been derided by them when it was new, and destroyed by them when they had the power.

They have fought every new truth ever heard of, and they have killed every truth-seeker who got into their hands.

II

The so-called religious organizations which now lead the war against the teaching of evolution are nothing more, at bottom, than conspiracies of the inferior man against his betters. They mirror very accurately his congenital hatred of knowledge, his bitter enmity to the man who knows more than he does, and so gets more out of life. Certainly it cannot have gone unnoticed that their membership is recruited, in the overwhelming main, from the lower orders—that no man of any education or other human dignity belongs to them. What they propose to do, at bottom and in brief, is to make the superior man infamous— by mere abuse if it is sufficient, and if it is not, then by law.

Such organizations, of course, must have leaders; there must be men in them whose ignorance and imbecility are measurably less abject than the ignorance and imbecility of the average. These super-Chandala often attain to a considerable power, especially in democratic states. Their followers trust them and look up to them; sometimes, when the pack is on the loose, it is necessary to conciliate them. But their puissance cannot conceal their incurable inferiority. They belong to the mob as surely as their dupes, and the thing that animates them is precisely the mob's hatred of superiority. Whatever lies above the level of their comprehension is of the devil. A glass of wine delights civilized men; they themselves, drinking it, would get drunk. *Ergo,* wine must be prohibited. The hypothesis of evolution is credited by all men of education; they themselves can't understand it. *Ergo,* its teaching must be put down.

This simple fact explains such phenomena as the Tennessee buffoonery. Nothing else can. We must think of human progress, not as of something going on in the race in general, but as of something going on in a small minority, perpetually beleaguered in a few walled towns. Now and then the horde of barbarians outside breaks through, and we have an armed effort to halt the process. That is, we have a Reformation, a French Revolution, a war for democracy, a Great Awakening. The minority is decimated and driven to cover. But a few survive—and a few are enough to carry on.

III

The inferior man's reasons for hating knowledge are not hard to discern. He hates it because it is complex—because it puts an unbearable burden upon his meager capacity for taking in ideas. Thus his search is always for short cuts. All superstitions are such short cuts. Their aim is to make the unintelligible simple, and even obvious. So on what seem to be higher levels, no man who has not had a long and arduous education can understand even the most elementary concepts of modern pathology. But even a hind at the plow can grasp the theory of chiropractic in two lessons. Hence the vast popularity of chiropractic among the submerged—and of osteopathy, Christian Science and other such quackeries with it. They are idiotic, but they are simple—and every man prefers what he can understand to what puzzles and dismays him.

The popularity of Fundamentalism among the inferior orders of men is explicable in exactly the same way. The cosmogonies that educated men toy with are all inordinately complex. To comprehend their veriest outlines requires an immense stock of knowledge, and a habit of thought. It would be as vain to try to teach to peasants or to the city proletariat as it would be to try to teach them to streptococci. But the cosmogony of Genesis is so simple that even a yokel can grasp it. It is set forth in a few phrases. It offers, to an ignorant man, the irresistible reasonableness of the nonsensical. So he accepts it with loud hosannas, and has one more excuse for hating his betters.

Politics and the fine arts repeat the story. The issues that the former throw up are often so complex that, in the present state of human knowledge, they must remain impenetrable, even to the most enlightened men. How much easier to follow a mountebank with a shibboleth— a Coolidge, a Wilson or a Roosevelt! The arts, like the sciences, demand special training, often very difficult. But in jazz there are simple rhythms, comprehensible even to savages.

IV

What all this amounts to is that the human race is divided into two sharply differentiated and mutually antagonistic classes, almost two genera—a small minority that plays with ideas and is capable of taking them in, and a vast majority that finds them painful, and is thus arrayed against them, and against all who have traffic with them. The intellectual heritage of the race belongs to the minority, and to the minority only. The majority has no more to do with it than it has to do with ecclesiastic politics on Mars. In so far as that heritage is apprehended, it is viewed with enmity. But in the main it is not apprehended at all.

That is why Beethoven survives. Of the 110,000,000 so-called human beings who now live in the United States, flogged and crazed by Coolidge, Rotary, the Ku Klux and the newspapers, it is probable that at least 108,000,000 have never heard of him at all. To these immortals, made in God's image, one of the greatest artists the human race has ever produced is not even a name. So far as they are concerned he might as well have died at birth. The gorgeous and incomparable beauties that he created are nothing to them. They get no value out of the fact that he existed. They are completely unaware of what he did in the world, and would not be interested if they were told.

The fact saves good Ludwig's bacon. His music survives because it lies outside the plane of the popular apprehension, like the colors beyond violet or the concept of honor. If it could be brought within range, it would at once arouse hostility. Its complexity would challenge; its lace of moral purpose would affright. Soon there would be a movement to put it down, and Baptist clergymen would range the land denouncing it, and in the end some poor musician, taken in the un-American act of playing it, would be put on trial before a jury of Ku Kluxers, and railroaded to the calaboose.

III

In Tennessee

From *The Nation*, July 1, 1925

Always, in this great republic, controversies depart swiftly from their original terms and plunge into irrelevancies and false pretenses. The case of prohibition is salient. Who recalls the optimistic days before the Eighteenth Amendment, and the lofty prognostication of the dry mullahs, clerical and lay? Prohibition, we were told, would empty the jails, reduce the tax rate, abolish poverty and put an end to political corruption. Today even the Prohibitionists know better, and so they begin to grow discreetly silent upon the matter. Instead, they come forward with an entirely new Holy Cause. What began as a campaign for a Babbitt's Utopia becomes transformed into a mystical campaign for Law Enforcement.

Prohibition is a grotesque failure, but the fight must go on. A transcendental motive takes the place of a practical motive. One categorical imperative goes out and another comes in.

So, now, in Tennessee, where a rural pedagogue stands arraigned before his peers for violating the school law. At bottom, a quite simple business. The hinds of the State, desiring to prepare their young for life there, set up public schools. To man those schools they employ pedagogues. To guide those pedagogues they lay down rules prescribing what is to be taught and what is not to be taught. Why not, indeed? How could it be otherwise? Precisely the same custom prevails everywhere else in the world, where there are schools at all. Behind every school ever heard of there is a definite concept of its purpose—of the sort of equipment it is to give to its pupils. It cannot conceivably teach everything; it must confine itself by sheer necessity to teaching what will be of the greatest utility, cultural or practical, to the youth actually in hand. Well, what could be of greater utility to the son of a Tennessee mountaineer than an education making him a good Tennesseean, content with his father, at peace with his neighbors, dutiful to the local religion, and docile under the local mores?

That is all the Tennessee anti-evolution law seeks to accomplish. It differs from other regulations of the same sort only to the extent that Tennessee differs from the rest of the world. The State, to a degree that should be gratifying, has escaped the national standardization. Its people show a character that is immensely different from the character of, say, New Yorkers or Californians. they retain, among other things, the anthropomorphic religion of an elder day. They do not profess it; they actually believe in it. the Old Testament, to them, is not a mere sacerdotal whiz-bang, to be read for its pornography; it is an authoritative history, and the transactions recorded in it are as true as the story of Barbara Frietchie,* or that of Washington and the cherry tree, or that of the late Woodrow's struggle to keep us out of the war. So crediting the sacred narrative, they desire that it be taught to their children, and any doctrine that makes game of it is immensely offensive to them. When such a doctrine, despite their protests, is actually taught, they proceed to put it down by force.

Is that procedure singular? I don't think it is. it is adopted everywhere, the instant the prevailing notions, whether real or false, are challenged. Suppose a school teacher in New York began entertaining his pupils with the case against the Jews, or against the Pope. Suppose a teacher in Vermont essayed to argue that the late Confederate States were right, as thousands of perfectly

* Nonagenarian flag-waving Civil War matron idealized in a John Whittier Greenleaf poem.

sane and intelligent persons believe—that Lee was a defender of the Constitution and Grant a traitor to it. Suppose a teacher in Kansas taught that prohibition was evil, or a teacher in New Jersey that it was virtuous. But I need not pile up suppositions. The evidence of what happens to such a contumacious teacher was spread before us copiously during the late uproar about Bolsheviks. And it was not in rural Tennessee but in the great cultural centers which now laugh at Tennessee that punishments came most swiftly, and were most barbarous. it was not Dayton but New York City that cashiered teachers for protesting against the obvious lies of the State Department.

Yet now we are asked to believe that some mysterious and vastly important principle is at stake at Dayton—that the conviction of Professor Scopes will strike a deadly blow at enlightenment and bring down freedom to sorrow and shame. Tell it to the marines! No principle is at stake at Dayton save the principle that school teachers, like plumbers, should stick to the job that is set before them, and not go roving about the house, breaking windows, raiding the cellar, and demoralizing the children. The issue of free speech is quite irrelevant. When a pedagogue takes his oath of office, he renounces his right to free speech quite as certainly as a bishop does, or a colonel in the army, or an editorial writer on a newspaper. He becomes a paid propagandist of certain definite doctrines and attitudes, mainly determined specifically and in advance, and every time he departs from them deliberately he deliberately swindles his employers.

What ails Mr. Scopes, and many like him, is that they have been filled with subversive ideas by specialists in human liberty, of whom I have the honor to be one. Such specialists, confronted by the New York cases, saw a chance to make political capital out of them, and did so with great effect. I was certainly not backward in that enterprise. The liars of the State Department were fair game, and any stick is good enough to beat a dog with. Even a pedagogue, seized firmly by the legs, makes an effective shillelagh. (I have used, in my time, yet worse: a congressman, a psychiatrist, a birth controller to maul an archbishop.) Unluckily, some of the pedagogues mistook the purpose of the operation. They came out of it full of a delusion that they were apostles of liberty, of the search for knowledge, of enlightenment. They have been worrying and exasperating their employers ever since.

I believe it must be plain that they are wrong, and that their employers, by a necessary inference, are tight. A pedagogue, properly so called—and a high-school teacher in a country town is properly so called—is surely not a searcher for knowledge. His job in the world is simply to pass on what has been chosen and approved by his superiors. In the whole history of the world no such pedagogue has ever actually increased the sum of human knowledge. His training unfits him for it; moreover, he would not be a pedagogue if he had either the taste or the capacity for it. He is a workingman, not a thinker. When he speaks, his employers speak. What he says has behind it all the authority of the community. If he would be true to his oath he must be very careful to say nothing that is in violation of the communal mores, the communal magic, the communal notion of the good, the beautiful, and the true.

Here, I repeat, I speak of the pedagogue, and use the word in its strict sense—that is, I speak of the fellow whose sole job is teaching. Men of great learning, men who genuinely know something, men who have augmented the store of human knowledge—such men, in their leisure, may also teach. The master may take an apprentice. But he does not seek apprentices in the hill towns of Tennessee, or even on the East Side of New York. He does not waste himself upon children whose fate it will be, when they grow up, to become Rotarians or Methodist deacons, bootleggers or moonshiners. He looks for his apprentices in the minority that has somehow escaped that fate—that has, by some act of God, survived the dreadful ministrations of school-teachers. To this minority he may submit his doubts as well as his certainties. He may present what is dubious and of evil report along with what is official, and hence good. He may be wholly himself. Liberty of teaching begins where pedagogy ends.

IV

Mencken Finds Daytonians Full of Sickening Doubts About Value of Publicity

From *The Baltimore Evening Sun*, July 9, 1925

Dayton, Tenn., July 9—On the eve of the great contest Dayton is full of sickening surges and tremors of doubt. Five or six weeks ago, when the infidel Scopes was first laid by the heels, there was no uncertainty in all this smiling valley. The town boomers leaped to the assault as one man. Here was an unexampled, almost a miraculous chance to get Dayton upon the front pages, to make it talked about, to put it upon the map. But how now?

Today, with the curtain barely rung up and the worst buffooneries to come, it is obvious to even town boomers that getting upon the map, like patriotism, is not enough. The getting there must be managed discreetly, adroitly, with careful regard to psychological niceties. The boomers of Dayton, alas, had no skill at such things, and the experts they called in were all quacks. The result now turns the communal liver to water. Two months ago the town was obscure and happy. Today it is a universal joke.

I have been attending the permanent town meeting that goes on in Robinson's drug store, trying to find out what the town optimists have saved from the wreck. All I can find is a sort of mystical confidence that God will somehow come to the rescue to reward His old and faithful partisans as they deserve—that good will flow eventually out of what now seems to be heavily evil. More specifically, it is believed that settlers will be attracted to the town as to some refuge from the atheism of the great urban Sodoms and Gomorrahs.

But will these refugees bring any money with them? Will they buy lots and build houses, Will they light the fires of the cold and silent blast furnace down the railroad tracks? On these points, I regret to report, optimism has to call in theology to aid it. Prayer can accomplish a lot. It can cure diabetes, find lost pocketbooks and restrain husbands from beating their wives. But is prayer made any more efficacious by giving a circus first? Coming to this thought, Dayton begins to sweat.

The town, I confess, greatly surprised me. I expected to find a squalid Southern village, with darkies snoozing on the horse-blocks, pigs rooting under the houses and the inhabitants full of hookworm and malaria. What I found was a country town full of charm and even beauty—a somewhat smallish but nevertheless very attractive Westminster or Belair.*

The houses are surrounded by pretty gardens, with cool green lawns and stately trees. The two chief streets are paved from curb to curb. The stores carry good stocks and have a metropolitan air, especially the drug, book, magazine, sporting goods and soda-water emporium of the estimable Robinson. A few of the town ancients still affect galluses and string ties, but the younger bucks are very nattily turned out. Scopes himself, even in his shirt sleeves, would fit into any college campus in America save that of Harvard alone.

Nor is there any evidence in the town of that poisonous spirit which usually shows itself when Christian men gather to defend the great doctrine of their faith. I have heard absolutely no whisper that Scopes is in the pay of the Jesuits, or that the whisky trust is backing him, or that he is egged on by the Jews who manufacture lascivious moving pictures. On the contrary, the Evolutionists and the Anti-Evolutionists seem to be on the best of terms, and it is hard in a group to distinguish one from another.

* Two prosperous towns in Maryland.

The basic issues of the case, indeed, seem to be very little discussed at Dayton. What interests everyone is its mere strategy. By what device, precisely, will Bryan trim old Clarence Darrow? Will he do it gently and with every delicacy of forensics, or will he wade in on high gear and make a swift butchery of it? For no one here seems to doubt that Bryan will win—that is, if the bout goes to a finish. What worries the town is the fear that some diabolical higher power will intervene on Darrow's side— that is, before Bryan heaves him through the ropes.

The lack of Christian heat that I have mentioned is probably due in part to the fact that the fundamentalists are in overwhelming majority as far as the eye can reach— according to most local statisticians, in a majority of at least nine-tenths. There are, in fact, only two downright infidels in all Rhea county, and one of them is charitably assumed to be a bit balmy. The other, a yokel roosting far back in the hills, is probably simply a poet got into the wrong pew. The town account of him is to the effect that he professes to regard death as a beautiful adventure.

When the local ecclesiastics begin alarming the peas- antry with word pictures of the last sad scene, and sulphurous fumes begin to choke even Unitarians, this skeptical rustic comes forward with his argument that it is foolish to be afraid of what one knows so little about—that, after all, there is no more genuine evidence that anyone will ever go to hell than there is that the Volstead act will ever be enforced.

Such blasphemous ideas naturally cause talk in a Baptist community, but both of the infidels are unmolested. Rhea county, in fact, is proud of its tolerance, and apparently with good reason. The Klan has never got a foothold here, though it rages everywhere else in Tennessee. When the first Kleagles came in they got the cold shoulder, and pretty soon they gave up the county as hopeless. It is run today not by anonymous daredevils in white nightshirts, but by well-heeled Free-masons in decorous white aprons. In Dayton alone there are sixty thirty-second-degree Masons— an immense quota for so small a town. They believe in keeping the peace, and so even the stray Catholics of the town are treated politely, though everyone naturally regrets they are required to report to the Pope once a week.

It is probably this unusual tolerance, and not any extraordinary passion for the integrity of Genesis, that has made Dayton the scene of a celebrated case, and got its name upon the front pages, and caused its forward-looking men to begin to wonder uneasily if all advertising is really good advertising. The trial of Scopes is possible here simply because it can be carried on here without heat—because no one will lose any sleep even if the devil comes to the aid of Darrow and Malone, and Bryan gets a mauling. The local intelligentsia venerate Bryan as a Christian, but it was not as a Christian that they called him in, but as one adept at attracting the newspaper boys—in brief, as a showman. As I have said, they now begin to mistrust the show, but they still believe that he will make a good one, win or lose.

Elsewhere, North or South, the combat would become bitter. Here it retains the lofty qualities of the *duello*. I gather the notion, indeed, that the gentlemen who are most active in promoting it are precisely the most lacking in hot conviction—that it is, in its local aspects, rather a joust between neutrals than a battle between passionate believers. Is it a mere coincidence that the town clergy have been very carefully kept out of it? There are several Baptist brothers here of such powerful gifts that when they begin belaboring sinners the very rats of the alleys flee to the hills. They preach dreadfully. But they are not heard from today. By some process to me unknown they have been induced to shut up—a far harder business, I venture, than knocking out a lion with a sandbag. But the sixty thirty-second degree Masons of Dayton have somehow achieved it.

Thus the battle joins and the good red sun shines down. Dayton lies in a fat and luxuriant valley. The bottoms are green with corn, pumpkins and young orchards and the hills are full of reliable moonshiners, all save one of them Christian men. We are not in the South here, but hanging on to the North. Very little cotton is grown in the valley. The people in politics are Republicans and put Coolidge next to Lincoln and John Wesley. The fences are in good repair. The roads are smooth and hard. The scene is set for a high-toned and even somewhat swagger combat. When it is over all the participants save Bryan will shake hands.

V

Impossibility of Obtaining Fair Jury Insures Scopes' Conviction, Says Mencken

From *The Baltimore Evening Sun*, July 10, 1925

Dayton, Tenn., July 10—The trial of the infidel Scopes, beginning here this hot, lovely morning, will greatly resemble, I suspect, the trial of a prohibition agent accused of mayhem in Union Hill, N.J. That is to say, it will be conducted with the most austere regard for the highest principles of jurisprudence. Judge and jury will go to extreme lengths to assure the prisoner the last and least of his rights. He will be protected in his person and feelings by the full military and naval power of the State of Tennessee. No one will be permitted to pull his nose, to pray publicly for his condemnation or even to make a face at him. But all the same he will be bumped off inevitably when the time comes, and to the applause of all right-thinking men.

The real trial, in truth, will not begin until Scopes is convicted and ordered to the hulks. Then the prisoner will be the Legislature of Tennessee, and the jury will be that great fair, unimpassioned body of enlightened men which has already decided that a horse hair put into a bottle will turn into a snake and that the Kaiser started the late war. What goes on here is simply a sort of preliminary hearing, with music by the village choir. For it will be no more possible in this Christian valley to get a jury unprejudiced against Scopes than would be possible in Wall Street to get a jury unprejudiced against a Bolshevik.

I speak of prejudice in its purely philosophical sense. As I wrote yesterday, there is an almost complete absence, in these pious hills, of the ordinary and familiar malignancy of Christian men. If the Rev. Dr. Crabbe ever spoke of bootleggers as humanely and affectionately as the town theologians speak of Scopes, and even Darrow and Malone, his employers would pelt him with their spyglasses and sit on him until the ambulance came from Mount Hope. There is absolutely no bitterness on tap. But neither is there any doubt. It has been decided by acclamation, with only a few infidels dissenting, that the hypothesis of evolution is profane, inhumane and against God, and all that remains is to translate that almost unanimous decision into the jargon of the law and so have done.

The town boomers have banqueted Darrow as well as Bryan, but there is no mistaking which of the two has the crowd, which means the venire of tried and true men. Bryan has been oozing around the country since his first day here, addressing this organization and that, presenting the indubitable Word of God in his caressing, ingratiating way, and so making unanimity doubly unanimous. From the defense yesterday came hints that this was making hay before the sun had legally begun to shine—even that it was a sort of contempt of court. But no Daytonian believes anything of the sort. What Bryan says doesn't seem to these congenial Baptists and Methodists to be argument; it seems to be a mere graceful statement of the obvious.

Meanwhile, reinforcements continue to come in, some of them from unexpected sources. I had the honor of being present yesterday when Col. Patrick Callahan, of Louisville, marched up at the head of his cohort of 250,000,000 Catholic fundamentalists. The two colonels embraced, exchanged a few military and legal pleasantries and then retired up a steep stairway to the office of the Hicks brothers to discuss strategy. Colonel Callahan's followers were present, of course, only by a legal fiction; the town of Dayton would not hold so large an army. In the actual flesh there were only the colonel himself and his aide-de-camp. Nevertheless, the 250,000,000 were put down as present and recorded as voting.

Later on I had the misfortune to fall into a dispute with Colonel Callahan on a point of canon law. It was my contention that the position of the Roman Church, on matters of doctrine, is not ordinarily stated by laymen— that such matters are usually left to high ecclesiastical authorities, headed by the Bishop of Rome. I also contended, perhaps somewhat fatuously, that there seemed to be a considerable difference of opinion regarding organic evolution among these authorities—that it was possible to find in their writings both ingenious arguments for it and violent protests against it. All these objections Colonel Callahan waived away with a genial gesture. He was here, he said, to do what he could for the authority of the Sacred Scriptures and the aiding and comforting of his old friend, Bryan, and it was all one to him whether atheists yelled or not. Then he began to talk about prohibition, which he favors, and the germ theory of diseases, which he regards as bilge.

A somewhat more plausible volunteer has turned up in the person of Pastor T.T. Martin, of Blue Mountain, Miss. He has hired a room and stocked it with pamphlets bearing such titles as "Evolution a Menace," "Hell and the High Schools" and "God or Gorilla," and addresses connoisseurs of scientific fallacy every night on a lot behind the Courthouse. Pastor Martin, a handsome and amiable old gentleman with a great mop of snow-white hair, was a professor of science in a Baptist college for years, and has given profound study to the biological sections of the Old Testament.

He told me today that he regarded the food regulations in Leviticus as so sagacious that their framing must have been a sort of feat even for divinity. The flesh of the domestic hog, he said, is a rank poison as ordinarily prepared for the table, though it is probably harmless when smoked and salted, as in bacon. He said that his investigations had shown that seven and a half out of every thirteen cows are quite free of tuberculosis, but that twelve out of every thirteen hogs have it in an advanced and highly communicable form. The Jews, protected by their piety against devouring pork, are immune to the disease. In all history, he said, there is authentic record of but one Jew who died of tuberculosis.

The presence of Pastor Martin and Colonel Callahan has given renewed confidence to the prosecution. The former offers proof that men of science are, after all, not unanimously atheists, and the latter that there is no division between Christians in the face of the common enemy. But though such encouragements help, they are certainly not necessary. All they really supply is another layer of icing on the cake. Dayton will give Scopes a rigidly fair and impartial trial. All his Constitutional rights will be jealously safeguarded. The question whether he voted for or against Coolidge will not be permitted to intrude itself into the deliberations of the jury, or the gallant effort of Colonel Bryan to get at and establish the truth. He will be treated very politely. Dayton, indeed, is proud of him, as Sauk Center, Minn., is proud of Sinclair Lewis and Whittingham, Vt., of Brigham Young. But it is lucky for Scopes that sticking pins into Genesis is still only a misdemeanor in Tennessee, punishable by a simple fine, with no alternative of the knout, the stone pile or exile to the Dry Tortugas.

VI

Mencken Likens Trial to a Religious Orgy, with Defendant a Beelzebub

From *The Baltimore Evening Sun*, July 11, 1925

Chattanooga, Tenn., July 11—Life down here in the Cumberland mountains realizes almost perfectly the ideal of those righteous and devoted men, Dr. Howard A. Kelly, the Rev. Dr. W.W. Davis, the Hon. Richard H. Edmonds and the Hon. Henry S. Dulaney. That is to say, evangelical Christianity is one hundred per cent triumphant. There is, of course, a certain subterranean heresy, but it is so cowed that it is almost inarticulate, and at its worst it would pass for the strictest orthodoxy in such Sodoms of infidelity as Baltimore. It may seem fabulous, but it is a sober fact that a sound Episcopalian or even a Northern Methodist would be regarded as virtually an atheist in Dayton. Here the only genuine conflict is between true believers. Of a given text in Holy Writ one faction may say this thing and another that, but both agree unreservedly that the text itself is impeccable, and neither in the midst of the most violent disputation would venture to accuse the other of doubt.

To call a man a doubter in these parts is equal to accusing him of cannibalism. Even the infidel Scopes himself is not charged with any such infamy. What they say of him, at worst, is that he permitted himself to be used as a cat's paw by scoundrels eager to destroy the anti-evolution law for their own dark and hellish ends. There is, it appears, a conspiracy of scientists afoot. Their purpose is to break down religion, propagate immorality, and so reduce mankind to the level of the brutes. They are the sworn and sinister agents of Beelzebub, who yearns to conquer the world, and has his eye especially upon Tennessee. Scopes is thus an agent of Beelzebub once removed, but that is as far as any fair man goes in condemning him. He is young and yet full of folly. When the secular arm has done execution upon him, the pastors will tackle him and he will be saved.

The selection of a jury to try him, which went on all yesterday afternoon in the atmosphere of a blast furnace, showed to what extreme lengths the salvation of the local primates has been pushed. It was obvious after a few rounds that the jury would be unanimously hot for Genesis. The most that Mr. Darrow could hope for was to sneak in a few men bold enough to declare publicly that they would have to hear the evidence against Scopes before condemning him. The slightest sign of anything further brought forth a peremptory challenge from the State. Once a man was challenged without examination for simply admitting that he did not belong formally to any church. Another time a panel man who confessed that he was prejudiced against evolution got a hearty round of applause from the crowd.

The whole process quickly took on an air of strange unreality, at least to a stranger from heathen parts. The desire of the judge to be fair to the defense, and even polite and helpful, was obvious enough—in fact, he more than once stretched the local rules of procedure in order to give Darrow a hand. But it was equally obvious that the whole thing was resolving itself into the trial of a man by his sworn enemies. A local pastor led off with a prayer calling on God to put down heresy; the judge himself charged the grand jury to protect the schools against subversive ideas. And when the candidates for the petit jury came up Darrow had to pass fundamentalist after fundamentalist into the box—some of them glaring at him as if they expected him to go off with a sulphurous bang every time he mopped his bald head.

In brief this is a strictly Christian community, and such is its notion of fairness, justice and due process of law. Try to picture a town made up wholly of Dr. Crabbes and Dr. Kellys, and you will have a reasonably accurate image of it. Its people are simply unable to imagine a man who rejects the literal authority of the Bible. The most they can conjure up, straining until they are red in the face, is a man who is in error about the meaning of this or that text. Thus one accused of heresy among them is like one accused of boiling his grandmother to make soap in Maryland. He must resign himself to being tried by a jury wholly innocent of any suspicion of the crime he is charged with and unanimously convinced that it is infamous. Such a jury, in the legal sense, may be fair. That is, it may be willing to hear the evidence against him before bumping him off. But it would certainly be spitting into the eye of reason to call it impartial.

The trial, indeed, takes on, for all its legal forms, something of the air of a religious orgy. The applause of the crowd I have already mentioned. Judge Raulston rapped it down and threatened to clear the room if it was repeated, but he was quite unable to still its echoes under his very windows. The courthouse is surrounded by a large lawn, and it is peppered day and night with evangelists. One and all they are fundamentalists and their yells and bawlings fill the air with orthodoxy. I have listened to twenty of them and had private discourse with a dozen, and I have yet to find one who doubted so much as the typographical errors in Holy Writ. They dispute raucously and far into the night, but they begin and end on the common ground of complete faith. One of these holy men wears a sign on his back announcing that he is the Bible champion of the world. He told me today that he had studied the Bible four hours a day for thirty-three years, and that he had devised a plan of salvation that would save the worst sinner ever heard of, even a scientist, a theater actor or a pirate on the high seas, in forty days. This gentleman denounced the hard-shell Baptists as swindlers. He admitted freely that their sorcerers were powerful preachers and could save any ordinary man from sin, but he said that they were impotent against iniquity. The distinction is unknown to city theologians, but is as real down here as that between sanctification and salvation. The local experts, in fact, debate it daily. The Bible champion, just as I left him, was challenged by one such professor, and the two were still hard at it an hour later.

Most of the participants in such recondite combats, of course, are yokels from the hills, where no sound is heard after sundown save the roar of the catamount and the wailing of departed spirits, and a man thus has time to ponder the divine mysteries. But it is an amazing thing that the more polished classes also participate actively. The professor who challenged the Bible champion was indistinguishable, to the eye, from a bond salesman or city bootlegger. He had on a natty palm beach suit and a fashionable soft collar and he used excellent English. Obviously, he was one who had been through the local high school and perhaps a country college. Yet he was so far uncontaminated by infidelity that he stood in the hot sun for a whole hour debating a point that even bishops might be excused for dodging, winter as well as summer.

The Bible champion is matched and rivaled by whole herds of other metaphysicians, and all of them attract good houses and have to defend themselves against constant attack. The Seventh Day Adventists, the Campbellites,* the Holy Rollers and a dozen other occult sects have field agents on the ground. They follow the traveling judges through all this country. Everywhere they go, I am told, they find the natives ready to hear them and dispute with them. They find highly accomplished theologians in every village, but even in the county towns they never encounter a genuine skeptic. If a man has doubts in this immensely pious country, he keeps them to himself.

Dr. Kelly should come down here and see his dreams made real. He will find a people who not only accept the

---

* Nickname of the Campbellite Christian Church, also known as Disciples of Christ

Bible as an infallible handbook of history, geology, biology and celestial physics, but who also practice its moral precepts—at all events, up to the limit of human capacity. It would be hard to imagine a more moral town than Dayton. If it has any bootleggers, no visitor has heard of them. Ten minutes after I arrived a leading citizen offered me a drink made up half of white mule and half of coca cola, but he seems to have been simply indulging himself in a naughty gesture. No fancy woman has been seen in the town since the end of the McKinley administration. There is no gambling. There is no place to dance. The relatively wicked, when they would indulge themselves, go to Robinson's drug store and debate theology.

In a word, the new Jerusalem, the ideal of all soul savers and sin exterminators. Nine churches are scarcely enough for the 1,800 inhabitants: many of them go into the hills to shout and roll. A clergyman has the rank and authority of a major-general of artillery. A Sunday-school superintendent is believed to have the gift of prophecy. But what of life here? Is it more agreeable than in Babylon? I regret that I must have to report that it is not. The incessant clashing of theologians grows monotonous in a day and intolerable the day following. One longs for a merry laugh, a burst of happy music, the gurgle of a decent jug. Try a meal in the hotel; it is tasteless and swims in grease. Go to the drug store and call for refreshment: the boy will hand you almost automatically a beaker of Coca-Cola. Look at the magazine counter: a pile of *Saturday Evening Post*s two feet high. Examine the books: melodrama and cheap amour. Talk to a town magnifico; he knows nothing that is not in Genesis.

I propose that Dr. Kelly be sent here for sixty days, preferably in the heat of summer. He will return to Baltimore yelling for a carboy of pilsner and eager to master the saxophone. His soul perhaps will be lost, but he will be a merry and a happy man.

VII

Yearning Mountaineers' Souls Need Reconversion Nightly, Mencken Finds

From *The Baltimore Evening Sun*, July 13, 1925

Dayton, Tenn., July 13—There is a Unitarian clergyman here from New York, trying desperately to horn into the trial and execution of the infidel Scopes. He will fail. If Darrow ventured to put him on the stand the whole audience, led by the jury, would leap out of the courthouse windows, and take to the hills. Darrow himself, indeed, is as much as they can bear. The whisper that he is an atheist has been stilled by the bucolic make-up and by the public report that he has the gift of prophecy and can reconcile Genesis and evolution. Even so, there is ample space about him when he navigates the streets. The other day a newspaper woman was warned by her landlady to keep out of the courtroom when he was on his legs.

All the local sorcerers predict that a bolt from heaven will fetch him in the end. The night he arrived there was a violent storm, the town water turned brown, and horned cattle in the lowlands were afloat for hours. A woman back in the mountains gave birth to a child with hair four inches long, curiously bobbed in scallops.

The Book of Revelation has all the authority, in these theological uplands, of military orders in time of war. The people turn to it for light upon all their problems, spiritual and secular. If a text were found in it denouncing the Anti-Evolution law, then the Anti-Evolution law would become infamous overnight. But so far the exegetes who roar and snuffle in the town have found no such text. Instead they have found only blazing ratifications and reinforcements of Genesis. Darwin is the devil with seven tails and nine horns. Scopes, though he is disguised by flannel pantaloons and a Beta Theta Pi haircut, is the harlot of Babylon. Darrow is Beelzebub in person and Malone is the Crown Prince Friedrich Wilhelm.

I have hitherto hinted an Episcopalian down here in the Coca-Cola belt is regarded as an atheist. It sounds like one of the lies that journalists tell, but it is really an understatement of the facts. Even a Methodist, by Rhea county standards, is one a bit debauched by pride of intellect. It is the four Methodists on the jury who are expected to hold out for giving Scopes Christian burial after he is hanged. They all made it plain, when they were examined, that they were free-thinking and independent men, and not to be run amuck by the superstitions of the lowly. One actually confessed that he seldom read the Bible, though he hastened to add that he was familiar with its principles. The fellow had on a boiled shirt and a polka dot necktie. He sits somewhat apart. When Darrow withers to a cinder under the celestial blowpipe, this dubious Wesleyan, too, will lose a few hairs.

Even the Baptists no longer brew a medicine that is strong enough for the mountaineers. The sacrament of baptism by total immersion is over too quickly for them, and what follows offers nothing that they can get their teeth into. What they crave is a continuous experience of the divine power, an endless series of evidence that the true believer is a marked man, ever under the eye of God. It is not enough to go to a revival once a year or twice a year; there must be a revival every night. And it is not enough to accept the truth as a mere statement of indisputable and awful fact: it must be embraced ecstatically and orgiastically, to the accompaniment of loud shouts, dreadful heavings and gurglings, and dancing with arms and legs.

This craving is satisfied brilliantly by the gaudy practices of the Holy Rollers, and so the mountaineers are gradually gravitating toward the Holy Roller communion, or, as they prefer to call it, the Church of God. Gradually, perhaps, is not the word. They are actually going in by whole villages and townships. At the last count of noses there were 20,000 Holy Rollers in these hills. The next census, I have no doubt, will show many more. The cities of the lowlands, of course, still resist, and so do most of the county towns, including even Dayton, but once one steps off the State roads the howl of holiness is heard in the woods, and the yokels carry on an almost continuous orgy.

A foreigner in store clothes going out from Dayton must approach the sacred grove somewhat discreetly. It is not that the Holy Rollers, discovering him, would harm him; it is simply that they would shut down their boiling of the devil and flee into the forests. We left Dayton an hour after nightfall and parked our car in a wood a mile or so beyond the little hill village of Morgantown. Far off in a glade a flickering light was visible and out of the silence came a faint rumble of exhortation. We could scarcely distinguish the figure of the preacher; it was like looking down the tube of a dark field microscope. We got out of the car and sneaked along the edge of a mountain cornfield.

Presently we were near enough to see what was going on. From the great limb of a mighty oak hung a couple of crude torches of the sort that car inspectors thrust under Pullman cars when a train pulls in at night. In their light was a preacher, and for a while we could see no one else. He was an immensely tall and thin mountaineer in blue jeans, his collarless shirt open at the neck and his hair a tousled mop. As he preached he paced up and down under the smoking flambeaux and at each turn he thrust his arms into the air and yelled, "Glory to God!" We crept nearer in the shadow of the cornfield and began to hear more of his discourse. He was preaching on the day of judgment. The high kings of the earth, he roared, would all fall down and die; only the sanctified would stand up to receive the Lord God of Hosts. One of these kings he mentioned by name—the king of what he called Greece-y. The King of Greece-y, he said, was doomed to hell.

We went forward a few more yards and began to see the audience. It was seated on benches ranged round the preacher in a circle. Behind him sat a row of elders, men and women. In front were the younger folk. We kept on cautiously, and individuals rose out of the ghostly gloom. A young mother sat suckling her baby, rocking as the preacher paced up and down. Two scared little girls hugged each other, their pigtails down their backs. An immensely huge mountain woman, in a gingham dress cut in one piece, rolled on her heels at every "Glory to God." To one side, but half visible, was what appeared to be a bed. We found out afterward that two babies were asleep upon it.

The preacher stopped at last and there arose out of the darkness a woman with her hair pulled back into a little tight knot. She began so quietly that we couldn't hear what she said, but soon her voice rose resonantly and we could follow her. She was denouncing the reading of books. Some wandering book agent, it appeared, had come to her cabin and tried to sell her a specimen of his wares. She refused to touch it. Why, indeed, read a book? If what was in it was true then everything in it was already in the Bible. If it was false then reading it would imperil the soul. Her syllogism complete, she sat down.

There followed a hymn, led by a somewhat fat brother wearing silver-rimmed country spectacles. It droned on for half a dozen stanzas, and then the first speaker resumed the floor. He argued that the gift of tongues was real and that education was a snare. Once his children could read the Bible, he said, they had enough. Beyond lay only infidelity and damnation. Sin stalked the cities. Dayton itself was a Sodom. Even Morgantown had begun to forget God. He sat down, and the female aurochs in gingham got up.

She began quietly, but was soon leaping and roaring, and it was hard to follow her. Under cover of the turmoil we sneaked a bit closer. A couple of other discourses followed, and there were two or three hymns. Suddenly a change of mood began to make itself felt. The last hymn ran longer than the others and dropped gradually into a monotonous, unintelligible chant. The leader beat time with his book. The faithful broke out with exultations. When the singing ended there was a brief palaver that we could not hear and two of the men moved a bench into the circle of light directly under the flambeaux. Then a half-grown girl emerged from the darkness and threw herself upon it. We noticed with astonishment that she had bobbed hair. "This sister," said the leader, "has asked for prayers." We moved a bit closer. We could now see faces plainly and hear every word.

What followed quickly reached such heights of barbaric grotesquerie that it was hard to believe it real. At a signal all the faithful crowded up the bench and began to pray—not in unison but each for himself. At another they all fell on their knees, their arms over the penitent. The leader kneeled, facing us, his head alternately thrown back dramatically or buried in his hands. Words spouted from his lips like bullets from a machine gun—appeals to God to pull the penitent back out of hell, defiances of the powers and principalities of the air, a vast impassioned jargon of apocalyptic texts. Suddenly he rose to his feet, threw back his head and began to speak in tongues—blub-blub-blub, gurgle-gurgle-gurgle. His voice rose to a higher register. The climax was a shrill, inarticulate squawk, like that of a man throttled. He fell headlong across the pyramid of supplicants.

A comic scene? Somehow, no. The poor half wits were too horribly in earnest. It was like peeping through a knothole at the writhings of a people in pain. From the squirming and jabbering mass a young woman gradually detached herself—a woman not uncomely, with a pathetic home-made cap on her head. Her head jerked back, the veins of her neck swelled, and her fists went to her throat as if she were fighting for breath. She bent backward until she was like half of a hoop. Then she suddenly snapped forward. We caught a flash of the whites of her eyes. Presently her whole body began to be convulsed—great convulsions that began at the shoulders and ended at the hips. She would leap to her feet, thrust her arms in air and then hurl herself upon the heap. Her praying flattened out into a mere delirious caterwauling, like that of a tomcat on a petting party.

I describe the thing as a strict behaviorist. The lady's subjective sensations I leave to infidel pathologists. Whatever they were they were obviously contagious, for soon another damsel joined her, and then another and then a fourth. The last one had an extraordinary bad attack. She began with mild enough jerks of the head, but in a moment she was bounding all over the place, exactly like a chicken with its head cut off. Every time her head came up a stream of yells and barkings would issue out of it. Once she collided with a dark, undersized brother, hitherto silent and stolid. Contact with her set him off as if he had been kicked by a mule. He leaped into the air, threw back his head and began to gargle as if with a mouthful of BB shot. Then he loosened one tremendous stentorian sentence in the tongues and collapsed.

By this time the performers were quite oblivious to the profane universe. We left our hiding and came up to the little circle of light. We slipped into the vacant seats on one of the rickety benches. The heap of mourners was directly before us. They bounced into us as they cavorted. The smell that they radiated, sweating there in that obscene heap, half suffocated us. Not all of them, of course, did the thing in the grand manner. Some merely moaned and rolled their eyes. The female ox in gingham flung her great bulk on the ground and jabbered an unintelligible prayer. One of the men, in the intervals between fits, put on spectacles and read his Bible.

Beside me on the bench sat the young mother and her baby. She suckled it through the whole orgy, obviously fascinated by what was going on, but never venturing to take any hand in it. On the bed just outside the light two other babies slept peacefully. In the shadows, suddenly appearing and as suddenly going away, were vague figures, whether of believers or of scoffers I do not know. They seemed to come and go in couples. Now and then a couple at the ringside would step back and then vanish into the black night. After a while some came back. There was whispering outside the circle of vision. A couple of Fords lurched up in the wood road, cutting holes in the darkness with their lights. Once some one out of sight loosed a bray of laughter.

All this went on for an hour or so. The original penitent, by this time, was buried three deep beneath the heap. One caught a glimpse, now and then, of her yellow bobbed hair, but then she would vanish again. How she breathed down there I don't know; it was hard enough ten feet away, with a strong five-cent cigar to help. When the praying brothers would rise up for a bout with the tongues their faces were streaming with perspiration. The fat harridan in gingham sweated like a longshoreman. Her hair got loose and fell down over her face. She fanned herself with her skirt. A powerful old gal she was, equal in her day to obstetrics and a week's washing on the same morning, but this was worse than a week's washing. Finally, she fell into a heap, breathing in great, convulsive gasps.

We tired of it after a while and groped our way back to our automobile. When we got to Dayton, after 11 o'clock—an immensely late hour in these parts—the whole town was still gathered on the courthouse lawn, hanging upon the disputes of theologians. The Bible champion of the world had a crowd. The Seventh Day Adventist missionaries had a crowd. A volunteer from faraway Portland, Ore., made up exactly like Andy Gump, had another and larger crowd. Dayton was enjoying itself. All the usual rules were suspended and the curfew bell was locked up. The prophet Bryan, exhausted by his day's work for Revelation, was snoring in his bed up the road, but enough volunteers were still on watch to keep the battlements manned.

Such is human existence among the fundamentalists, where children are brought up on Genesis and sin is unknown. If I have made the tale too long, then blame the spirit of garrulity that is in the local air. Even newspaper reporters, down here, get some echo of the call. Divine inspiration is as common as the hookworm. I have done my best to show you what the great heritage of mankind comes to in regions where the Bible is the beginning and end of wisdom, and the mountebank Bryan, parading the streets in his seersucker coat, is pointed out to sucklings as the greatest man since Abraham.

VIII

Darrow's Eloquent Appeal Wasted on Ears That Heed Only Bryan, Says Mencken

From *The Baltimore Evening Sun*, July 14, 1925

Dayton, Tenn., July 14—The net effect of Clarence Darrow's great speech yesterday seems to be precisely the same as if he had bawled it up a rainspout in the interior of Afghanistan. That is, locally, upon the process against the infidel Scopes, upon the so-called minds of these fundamentalists of upland Tennessee. You have but a dim notion of it who have only read it. It was not designed for reading, but for hearing. The clanging of it was as important as the logic. It rose like a wind and ended like a flourish of bugles. The very judge on the bench, toward the end of it, began to look uneasy. But the morons in the audience, when it was over, simply hissed it.

During the whole time of its delivery the old mountebank, Bryan, sat tight-lipped and unmoved. There is, of course, no reason why it should have shaken him. He has those hillbillies locked up in his pen and he knows it. His brand is on them. He is at home among them. Since his earliest days, indeed, his chief strength has been among the folk of remote hills and forlorn and lonely farms. Now with his political aspirations all gone to pot, he turns to them for religious consolations. They understand his peculiar imbecilities. His nonsense is their ideal of sense. When he deluges them with his theological bilge they rejoice like pilgrims disporting in the river Jordan.

The town whisper is that the local attorney general, Stewart, is not a fundamentalist, and hence has no stomach for his job. It seems not improbable. He is a man of evident education, and his argument yesterday was confined very strictly to the constitutional points—the argument of a competent and conscientious lawyer, and to me, at least, very persuasive.

But Stewart, after all, is a foreigner here, almost as much so as Darrow or Hays or Malone. He is doing his job and that is all. The real animus of the prosecution centers in Bryan. He is the plaintiff and prosecutor. The local lawyers are simply bottle-holders for him. He will win the case, not by academic appeals to law and precedent, but by direct and powerful appeals to the immemorial fears and superstitions of man. It is no wonder that he is hot against Scopes. Five years of Scopes and even these mountaineers would begin to laugh at Bryan. Ten years and they would ride him out of town on a rail, with one Baptist parson in front of him and another behind.

But there will be no ten years of Scopes, nor five years, nor even one year.

Such brash young fellows, debauched by the enlightenment, must be disposed of before they become dangerous, and Bryan is here, with his tight lips and hard eyes, to see that this one is disposed of. The talk of the lawyers, even the magnificent talk of Darrow, is so much idle wind music. The case will not be decided by logic, nor even by eloquence. It will be decided by counting noses—and for every nose in these hills that has ever thrust itself into any book save the Bible there are a hundred adorned with the brass ring of Bryan. These are his people. They understand him when he speaks in tongues. The same dark face that is in his own eyes is in theirs, too. They feel with him, and they relish him.

I sincerely hope that the nobility and gentry of the lowlands will not make the colossal mistake of viewing this trial of Scopes as a trivial farce. Full of rustic japes and in bad taste, it is, to be sure, somewhat comic on the surface. One laughs to see lawyers sweat. The jury, marched down Broadway, would set New York by the ears. But all of that is only skin deep.

Deeper down there are the beginnings of a struggle that may go on to melodrama of the first caliber, and when the curtain falls at last all the laughter may be coming from the yokels. You probably laughed at the prohibitionists, say, back in 1914. Well, don't make the same error twice.

As I have said, Bryan understands these peasants, and they understand him. He is a bit mangey and flea-bitten, but by no means ready for his harp. He may last five years, ten years or even longer. What he may accomplish in that time, seen here at close range, looms up immensely larger than it appears to a city man five hundred miles away. The fellow is full of such bitter, implacable hatreds that they radiate from him like heat from a stove. He hates the learning that he cannot grasp. He hates those who sneer at him. He hates, in general, all who stand apart from his own pathetic commonness. And the yokels hate with him, some of them almost as bitterly as he does himself. They are willing and eager to follow him—and he has already given them a taste of blood.

Darrow's peroration yesterday was interrupted by Judge Raulston, but the force of it got into the air nevertheless. This year it is a misdemeanor for a country school teacher to flout the archaic nonsense of Genesis. Next year it will be a felony. The year after the net will be spread wider. Pedagogues, after all, are small game; there are larger birds to snare—larger and juicier. Bryan has his fishy eye on them. He will fetch them if his mind lasts, and the lamp holds out to burn. No man with a mouth like that ever lets go. Nor ever lacks followers.

Tennessee is bearing the brunt of the first attack simply because the civilized minority, down here, is extraordinarily pusillanimous.

I have met no educated man who is not ashamed of the ridicule that has fallen upon the State, and I have met none, save only Judge Neal, who had the courage to speak out while it was yet time. No Tennessee counsel of any importance came into the case until yesterday and then they came in stepping very softly as if taking a brief for sense were a dangerous matter. When Bryan did his first rampaging here all these men were silent.

They had known for years what was going on in the hills. They knew what the country preachers were preaching—what degraded nonsense was being rammed and hammered into yokel skulls. But they were afraid to go out against the imposture while it was in the making, and when any outsider denounced it they fell upon him violently as an enemy of Tennessee.

Now Tennessee is paying for that poltroonery. The State is smiling and beautiful, and of late it has begun to be rich. I know of no American city that is set in more lovely scenery than Chattanooga, or that has more charming homes. The civilized minority is as large here, I believe, as anywhere else.

It has made a city of splendid material comforts and kept it in order. But it has neglected in the past the unpleasant business of following what was going on in the crossroads Little Bethels.

The Baptist preachers ranted unchallenged.

Their buffooneries were mistaken for humor. Now the clowns turn out to be armed, and have begun to shoot.

In his argument yesterday Judge Neal had to admit pathetically that it was hopeless to fight for a repeal of the anti-evolution law. The Legislature of Tennessee, like the Legislature of every other American state, is made up of cheap job-seekers and ignoramuses.

The Governor of the State is a politician ten times cheaper and trashier. It is vain to look for relief from such men. If the State is to be saved at all, it must be saved by the courts. For one, I have little hope of relief in that direction, despite Hays' logic and Darrow's eloquence. Constitutions, in America, no longer mean what they say. To mention the Bill of Rights is to be damned as a Red.

The rabble is in the saddle, and down here it makes its first campaign under a general beside whom Wat Tylor* seems like a wart beside the Matterhorn.

* Leader of the 1381 Peasants' Revolt against King Richard II

IX

Law and Freedom, Mencken Discovers, Yield Place to Holy Writ in Rhea County

From *The Baltimore Evening Sun*, July 15, 1925

Dayton, Tenn., July 15—The cops have come up from Chattanooga to help save Dayton from the devil. Darrow, Malone and Hays, of course, are immune to constabulary process, despite their obscene attack upon prayer. But all other atheists and anarchists now have public notice they must shut up forthwith and stay shut so long as they pollute this bright, shining, buckle of the Bible belt with their presence. Only one avowed infidel has ventured to make a public address. The Chattanooga police nabbed him instantly, and he is now under surveillance in a hotel. Let him but drop one of his impious tracts from his window and he will be transferred to the town hoose-gow.

The Constitution of Tennessee, as everyone knows, puts free speech among the most sacred rights of the citizen. More, I am informed by eminent Chattanooga counsel that there is no State law denying it—that is, for persons not pedagogues. But the cops of Chattanooga, like their brethren elsewhere, do not let constitutions stand in the way of their exercise of their lawful duty. The captain in charge of the squad now on watch told me frankly yesterday that he was not going to let any infidels discharge their damnable nonsense upon the town. I asked him what charge he would lay against them if they flouted him. He said he would jail them for disturbing the peace.

"But suppose," I asked him, "a prisoner is actually not disturbing the peace. Suppose he is simply saying his say in a quiet and orderly manner."

"I'll arrest him anyhow," said the cop.

"Even if no one complains of him?"

"I'll complain myself."

"Under what law precisely?"

"We don't need no law for them kind of people."

It sounded like New York in the old days, before Mayor Gaynor took the constitution out of cold storage and began to belabor the gendarmerie with it. The captain admitted freely that speaking in the streets was not disturbing the peace so long as the speaker stuck to orthodox Christian doctrine as it is understood by the local exegetes.

A preacher of any sect that admits the literal authenticity of Genesis is free to gather a crowd at any time and talk all he wants. More, he may engage in a disputation with any other expert. I have heard at least a hundred such discussions, and some of them have been very acrimonious. But the instant a speaker utters a word against divine revelation he begins to disturb the peace and is liable to immediate arrest and confinement in the calaboose beside the railroad tracks.

Such is criminal law in Rhea county as interpreted by the uniformed and freely sweating agents. As I have said, there are legal authorities in Chattanooga who dissent sharply, and even argue that the cops are a set of numbskulls and ought to be locked up as public nuisances. But one need not live a long, incandescent week in the Bible belt to know that jurisprudence becomes a new science as one crosses the border. Here the ordinary statutes are reinforced by Holy Writ, and whenever there is a conflict Holy Writ takes precedence.

Judge Raulston himself has decided, in effect, that in a trial for heresy it is perfectly fair and proper to begin proceedings with a prayer for the confutation and salvation of the defendant. On lower levels, and especially in the depths where policemen do their thinking, the doctrine is even more frankly stated. Before laying Christians by the heels the cops must formulate definite charges against them.

They must be accused of something specifically unlawful and there must be witnesses to the act. But infidels are *fera naturae,* and any cop is free to bag at sight and to hold them in durance at his pleasure.

To the same category, it appears, belong political and economic radicals. News came the other day to Pastor T.T. Martin, who is holding a continuous anti-evolution convention in the town, that a party of I.W.W.'s, their pockets full of Russian gold, had started out from Cincinnati to assassinate him. A bit later came word they would bump off Bryan after they had finished Martin, and then set fire to the town churches. Martin first warned Bryan and then complained to the police. The latter were instantly agog. Guards were posted at strategic centers and a watch was kept upon all strangers of a sinister appearance. But the I.W.W.'s were not caught. Yesterday Pastor Martin told me that he had news that they had gone back to Cincinnati to perfect the plot. He posts audiences at every meeting. If the Reds return they will be scotched.

Arthur Garfield Hays, who is not only one of the counsel for the infidel Scopes but also agent and attorney of the notorious American Civil Liberties Union in New York, is planning to hold a free speech meeting on the Courthouse lawn and so make a test of the law against disturbing the peace as it is interpreted by the *Polizei.*

Hays will be well advertised if he carries out this subversive intention. It is hot enough in the courtroom in the glare of a thousand fundamentalist eyes; in the town jail he would sweat to death.

Rhea county is very hospitable and, judged by Bible belt standards, very tolerant. The Dayton Babbitts gave a banquet to Darrow, despite the danger from lightning, meteors and earthquakes. Even Malone is treated politely, though the very horned cattle in the fields know that he is a Catholic and in constant communication with the Pope. But liberty is one thing and license is quite another. Within the bounds of Genesis the utmost play of opinion is permitted and even encouraged. An evangelist with a new scheme for getting into Heaven can get a crowd in two minutes. But once a speaker admits a doubt, however cautiously, he is handed over to the secular arm.

Two Unitarian clergymen are prowling around the town looking for a chance to discharge their "hellish heresies." One of them is Potter, of New York; the other is Birckhead, of Kansas City.* So far they have not made any progress. Potter induced one of the local Methodist parsons to give him a hearing, but the congregation protested and the next day the parson had to resign his charge. The Methodists, as I have previously reported, are regarded almost as infidels in Rhea county. Their doctrines, which seem somewhat

---

* Charles Francis Potter, a clergyman of the West Side Unitarian Church, helped Darrow as a Bible expert.

severe in Baltimore, especially to persons who love a merry life, are here viewed as loose to the point of indecency. The four Methodists on the jury are suspected of being against hanging Scopes, at least without a fair trial. The State tried to get rid of one of them even after he had been passed; his neighbors had come in from his village with news that he had a banjo concealed in his house and was known to read the *Literary Digest*.

The other Unitarian clergyman, Dr. Birckhead, is not actually domiciled in the town, but is encamped, with his wife and child, on the road outside. He is on an automobile tour and stopped off here to see if a chance offered to spread his "poisons." So far he has found none.

Yesterday afternoon a Jewish rabbi from Nashville also showed up, Marks by name. He offered to read and expound Genesis in Hebrew, but found no takers. The Holy Rollers hereabout, when they are seized by the gift of tongues, avoid Hebrew, apparently as a result of Ku Klux influence. Their favorite among all the sacred dialects is Hittite. It sounds to the infidel like a series of college yells.

Judge Raulston's decision yesterday afternoon in the matter of Hays' motion was a masterpiece of unconscious humor. The press stand, in fact, thought he was trying to be jocose deliberately and let off a guffaw that might have gone far if the roar of applause had not choked it off.

Hays presented a petition in the name of the two Unitarians, the rabbi and several other theological "Reds," praying that in selecting clergymen to open the court with prayer hereafter he choose fundamentalists and anti-fundamentalists alternately. The petition was couched in terms that greatly shocked and enraged the prosecution. When the judge announced that he would leave the nomination of chaplains to the Pastors' Association of the town there was the gust of mirth aforesaid, followed by howls of approval. The Pastors' Association of Dayton is composed of fundamentalists so powerfully orthodox that beside them such a fellow as Dr. John Roach Straton would seem an Ingersoll.*

The witnesses of the defense, all of them heretics, began to reach town yesterday and are all quartered at what is called the Mansion, an ancient and empty house outside the town limits, now crudely furnished with iron cots, spittoons, playing cards and the other camp equipment of scientists. Few, if any, of these witnesses will ever get a chance to outrage the jury with their blasphemies, but they are of much interest to the townspeople. The common belief is that they will be blown up with one mighty blast when the verdict of the twelve men, tried and true, is brought in, and Darrow, Malone, Hays and Neal with them. The country people avoid the Mansion.

* Stratton was a principal leader of the anti-evolution campaign. He debated Potter extensively.

It is foolish to take unnecessary chances. Going into the courtroom, with Darrow standing there shamelessly and openly challenging the wrath of God, is risk enough.

The case promises to drag into next week. The prosecution is fighting desperately and taking every advantage of its superior knowledge of the quirks of local procedure. The defense is heating up and there are few exchanges of courtroom amenities. There will be a lot of oratory before it is all over and some loud and raucous bawling otherwise, and maybe more than one challenge to step outside. The cards seem to be stacked against poor Scopes, but there may be a joker in the pack. Four of the jurymen, as everyone knows, are Methodists, and a Methodist down here belongs to the extreme wing of liberals. Beyond him lie only the justly and incurably damned.

What if one of those Methodists, sweating under the dreadful pressure of fundamentalist influence, jumps into the air, cracks his heels together and gives a defiant yell? What if the jury is hung? It will be a good joke on the fundamentalists if it happens, and an even better joke on the defense.

X

Mencken Declares Strictly Fair Trial Is Beyond Ken of Tennessee Fundamentalists

From *The Baltimore Evening Sun*, July 16, 1925

Dayton, Tenn., July 16—Two things ought to be understood clearly by heathen Northerners who follow the great cause of the State of Tennessee against the infidel Scopes. One is that the old mountebank, Bryan, is no longer thought of as a mere politician and jobseeker in these Godly regions, but has become converted into a great sacerdotal figure, half man and half archangel—in brief, a sort of fundamentalist pope. The other is that the fundamentalist mind, running in a single rut for fifty years, is now quite unable to comprehend dissent from its basic superstitions, or to grant any common honesty, or even any decency, to those who reject them.

The latter fact explains some of the most astonishing singularities of the present trial—that is, singularities to one accustomed to more austere procedures. In the average Northern jurisdiction much of what is going on here would be almost unthinkable. Try to imagine a trial going on in a town in which anyone is free to denounce the defendant's case publicly and no one is free to argue for it in the same way—a trial in a courthouse placarded with handbills set up by his opponents—a trial before a jury of men who have been roweled and hammered by those opponents for years, and have never heard a clear and fair statement of his answer.

But this is not all. It seems impossible, but it is nevertheless a fact that public opinion in Dayton sees no impropriety in the fact that the case was opened with prayer by a clergyman known by everyone to be against Scopes and by no means shy about making the fact clear. Nor by the fact that Bryan, the actual complainant, has been preparing the ground for the prosecution for months. Nor by the fact that, though he is one of the attorneys of record in the case, he is also present in the character of a public evangelist and that throngs go to hear him whenever he speaks, including even the sitting judge.

I do not allege here that there is any disposition to resort to lynch law. On the contrary, I believe that there is every intent to give Scopes a fair trial, as a fair trial is understood among fundamentalists. All I desire to show is that all the primary assumptions are immovably against him—that it is a sheer impossibility for nine-tenths of those he faces to see any merit whatever in his position. He is not simply one who has committed a misdemeanor against the peace and dignity of the State, he is also the agent of a heresy almost too hellish to be stated by reputable men. Such reputable men recognize their lawful duty to treat him humanely and even politely, but they also recognize their superior duty to make it plain that they are against his heresy and believe absolutely in the wisdom and virtue of his prosecutors.

In view of the fact that everyone here looks for the jury to bring in a verdict of guilty, it might be expected that the prosecution would show a considerable amiability and allow the defense a rather free plan. Instead, it is contesting every point very vigorously and taking every advantage of its greatly superior familiarity with local procedure. There is, in fact, a considerable heat in the trial. Bryan and the local lawyers for the State sit glaring at the defense all day and even the Attorney General, A.T. Stewart, who is supposed to have secret doubts about fundamentalism, has shown such pugnacity that it has already brought him to forced apologies.

The high point of yesterday's proceedings was reached with the appearance of Dr. Maynard M. Metcalfe, of the

Johns Hopkins. The doctor is a somewhat chubby man of bland mien, and during the first part of his testimony, with the jury present, the prosecution apparently viewed him with great equanimity. But the instant he was asked a question bearing directly upon the case at bar there was a flurry in the Bryan pen and Stewart was on his feet with protests. Another question followed, with more and hotter protests. The judge then excluded the jury and the show began.

What ensued was, on the surface, a harmless enough dialogue between Dr. Metcalfe and Darrow, but underneath there was very tense drama. At the first question Bryan came out from behind the State's table and planted himself directly in front of Dr. Metcalfe, and not ten feet away. The two McKenzies followed, with young Sue Hicks at their heels.

Then began one of the clearest, most succinct and withal most eloquent presentations of the case for the evolutionists that I have ever heard. The doctor was never at a loss for a word, and his ideas flowed freely and smoothly. Darrow steered him magnificently. A word or two and he was howling down the wind. Another and he hauled up to discharge a broadside. There was no cocksureness in him. Instead he was rather cautious and deprecatory and sometimes he halted and confessed his ignorance. But what he got over before he finished was a superb counterblast to the fundamentalist buncombe.

The jury, at least, in theory heard nothing of it, but it went whooping into the radio and it went banging into the face of Bryan.

Bryan sat silent throughout the whole scene, his gaze fixed immovably on the witness. Now and then his face darkened and his eyes flashed, but he never uttered a sound. It was, to him, a string of blasphemies out of the devil's mass—a dreadful series of assaults upon the only true religion. The old gladiator faced his real enemy at last. Here was a sworn agent and attorney of the science he hates and fears—a well-fed, well-mannered spokesman of the knowledge he abominates. Somehow he reminded me pathetically of the old Holy Roller I heard last week— the mountain pastor who damned education as a mocking and a corruption. Bryan, too, is afraid of it, for wherever it spreads his trade begins to fall off, and wherever it flourishes he is only a poor clown.

But not to these fundamentalists of the hills. Not to yokels he now turns to for consolation in his old age, with the scars of defeat and disaster all over him. To these simple folk, as I have said, he is a prophet of the imperial line—a lineal successor to Moses and Abraham. The barbaric cosmogony that he believes in seems as reasonable to them as it does to him. They share his peasant-like suspicion of all book learning that a plow hand cannot grasp. They believe with him that men who know too much should be seized by the secular arm and put down by force. They dream as he does of a world unanimously sure of Heaven and unanimously idiotic on this earth.

This old buzzard, having failed to raise the mob against its rulers, now prepares to raise it against its teachers. He can never be the peasants' President, but there is still a chance to be the peasants' Pope. He leads a new crusade, his bald head glistening, his face streaming with sweat, his chest heaving beneath his rumpled alpaca coat. One somehow pities him, despite his so palpable imbecilities. It is a tragedy, indeed, to begin life as a hero and to end it as a buffoon. But let no one, laughing at him, underestimate the magic that lies in his black, malignant eye, his frayed but still eloquent voice. He can shake and inflame these poor ignoramuses as no other man among us can shake and inflame them, and he is desperately eager to order the charge.

In Tennessee he is drilling his army. The big battles, he believes, will be fought elsewhere.

XI

Malone the Victor, Even Though Court Sides with Opponents, Says Mencken

From *The Baltimore Evening Sun*, July 17, 1925

Dayton, Tenn., July 17—Though the court decided against him this morning, and the testimony of the experts summoned for the defense will be banned out of the trial of the infidel Scopes, it was Dudley Field Malone who won yesterday's great battle of rhetoricians. When he got upon his legs it was the universal assumption in the courtroom that Judge Raulston's mind was already made up, and that nothing that any lawyer for the defense could say would shake him. But Malone unquestionably shook him. He was, at the end, in plain doubt, and he showed it by his questions. It took a night's repose to restore him to normalcy. The prosecution won, but it came within an inch of losing.

Malone was put up to follow and dispose of Bryan, and he achieved the business magnificently. I doubt that any louder speech has ever been heard in a court of law since the days of Gog and Magog. It roared out of the open windows like the sound of artillery practice, and alarmed the moonshiners and catamounts on distant peaks. Trains thundering by on the nearby railroad sounded faint and far away and when, toward the end, a table covered with standing and gaping journalists gave way with a crash, the noise seemed, by contrast, to be no more than a *pizzicato* chord upon a viola da gamba. The yokels outside stuffed their Bibles into the loud-speaker horns and yielded themselves joyously to the impact of the original. In brief, Malone was in good voice. It was a great day for Ireland. And for the defense. For Malone not only out-yelled Bryan, he also plainly out-generaled and out-argued him. His speech, indeed, was one of the best presentations of the case against the fundamentalist rubbish that I have ever heard.

It was simple in structure, it was clear in reasoning, and at its high points it was overwhelmingly eloquent. It was not long, but it covered the whole ground and it let off many a gaudy skyrocket, and so it conquered even the fundamentalists. At its end they gave it a tremendous cheer—a cheer at least four times as hearty as that given to Bryan. For these rustics delight in speechifying, and know when it is good. The devil's logic cannot fetch them, but they are not above taking a voluptuous pleasure in his lascivious phrases.

The whole speech was addressed to Bryan, and he sat through it in his usual posture, with his palm-leaf fan flapping energetically and his hard, cruel mouth shut tight. The old boy grows more and more pathetic. He has aged greatly during the past few years and begins to look elderly and enfeebled. All that remains of his old fire is now in his black eyes. They glitter like dark gems, and in their glitter there is immense and yet futile malignancy. That is all that is left of the Peerless Leader of thirty years ago. Once he had one leg in the White House and the nation trembled under his roars. Now he is a tinpot pope in the Coca-Cola belt and a brother to the forlorn pastors who belabor half-wits in galvanized iron tabernacles behind the railroad yards. His own speech was a grotesque performance and downright touching in its imbecility. Its climax came when he launched into a furious denunciation of the doctrine that man is a mammal. It seemed a sheer impossibility that any literate man should stand up in public and discharge any such nonsense. Yet the poor old fellow did it. Darrow stared incredulous. Malone sat with his mouth wide open. Hays indulged himself one of his sardonic chuckles. Stewart and Bryan fils looked extremely uneasy, but the old mountebank ranted on. To call a man a mammal, it appeared, was to flout the revelation of God. The certain effect of the doctrine would be to destroy morality and promote infidelity. The defense let it pass. The lily needed no gilding.

There followed some ranting about the Leopold-Loeb case, culminating in the argument that learning was corrupting—that the colleges by setting science above Genesis were turning their students into murderers. Bryan alleged that Darrow had admitted the fact in his closing speech at the Leopold-Loeb trial, and stopped to search for the passage in a printed copy of the speech. Darrow denied making any such statement, and presently began reading what he actually had said on the subject. Bryan then proceeded to denounce Nietzsche, whom he described as an admirer and follower of Darwin. Darrow challenged the fact and offered to expound what Nietzsche really taught. Bryan waved him off.

The effect of the whole harangue was extremely depressing. It quickly ceased to be an argument addressed to the court—Bryan, in fact, constantly said "My friends" instead of "Your Honor"—and became a sermon at the camp-meeting. All the familiar contentions of the Dayton divines appeared in it—that learning is dangerous, that nothing is true that is not in the Bible, that a yokel who goes to church regularly knows more than any scientist ever heard of. The thing went to fantastic lengths.

It became a farrago of puerilities without coherence or sense. I don't think the old man did himself justice. He was in poor voice and his mind seemed to wander. There was far too much hatred in him for him to be persuasive.

The crowd, of course, was with him. It has been fed upon just such balderdash for years. Its pastors assault it twice a week with precisely the same nonsense. It is chronically in the position of a populace protected by an espionage act in time of war. That is to say, it is forbidden to laugh at the arguments of one side and forbidden to hear the case of the other side. Bryan has been roving around in the tall grass for years and he knows the bucolic mind. He knows how to reach and inflame its basic delusions and superstitions. He has taken them into his own stock and adorned them with fresh absurdities. Today he may well stand as the archetype of the American rustic. His theology is simply the elemental magic that is preached in a hundred thousand rural churches fifty-two times a year.

These Tennessee mountaineers are not more stupid than the city proletariat; they are only less informed. If Darrow, Malone and Hays could make a month's stumping tour in Rhea county I believe that fully a fourth of the population would repudiate fundamentalism, and that not a few of the clergy now in practice would be restored to their old jobs on the railroad. Malone's speech yesterday probably shook a great many true believers; another like it would fetch more than one of them. But the chances are heavily against them ever hearing a second. Once this trial is over, the darkness will close in again, and it will take long years of diligent and thankless effort to dispel it—if, indeed, it is ever dispelled at all.

With a few brilliant exceptions—Dr. Neal is an example—the more civilized Tennesseeans show few signs of being equal to the job. I suspect that politics is what keeps them silent and makes their State ridiculous. Most of them seem to be candidates for office, and a candidate for office, if he would get the votes of fundamentalists, must bawl for Genesis before he begins to bawl for anything else. A typical Tennessee politician is the Governor, Austin Peay. He signed the anti-evolution bill with loud hosannas, and he is now making every effort to turn the excitement of the Scopes trial to his private political uses. The local papers print a telegram that he has sent to Attorney General A.T. Stewart whooping for prayer. In the North a Governor who indulged in such monkey shines would be rebuked for trying to influence the conduct of a case in court. And he would be derided as a cheap mountebank. But not here.

I described Stewart the other day as a man of apparent education and sense and palpably superior to the village lawyers who sit with him at the trial table. I still believe that I described him accurately. Yet even Stewart toward the close of yesterday's session gave an exhibition that would be almost unimaginable in the North. He began his reply to Malone with an intelligent and forceful legal argument, with plenty of evidence of hard study in it. But presently he slid into a violent theological harangue, full of extravagant nonsense. He described the case as a combat between light and darkness and almost descended to the depths of Bryan. Hays challenged him with a question. Didn't he admit, after all, that the defense had a tolerable case; that it ought to be given a chance to present its evidence? I transcribe his reply literally:

"That which strikes at the very foundations of Christianity is not entitled to a chance."

Hays, plainly astounded by this bald statement of the fundamentalist view of due process, pressed the point. Assuming that the defense would present, not opinion but only unadorned fact, would Stewart still object to its admission? He replied.

"Personally, yes."

"But as a lawyer and attorney general?" insisted Hays.

"As a lawyer and attorney general," said Stewart, "I am the same man."

Such is justice where Genesis is the first and greatest of law books and heresy is still a crime.

XII

Battle Now Over, Mencken Sees; Genesis Triumphant and Ready for New Jousts

From *The Baltimore Evening Sun*, July 18, 1925

Dayton, Tenn., July 18—All that remains of the great cause of the State of Tennessee against the infidel Scopes is the formal business of bumping off the defendant. There may be some legal jousting on Monday and some gaudy oratory on Tuesday, but the main battle is over, with Genesis completely triumphant. Judge Raulston finished the benign business yesterday morning by leaping with soft judicial hosannas into the arms of the prosecution. The sole commentary of the sardonic Darrow consisted of bringing down a metaphorical custard pie upon the occiput of the learned jurist.

"I hope," said the latter nervously, "that counsel intends no reflection upon this court."

Darrow hunched his shoulders and looked out of the window dreamily.

"Your honor," he said, "is, of course, entitled to hope."

No doubt the case will be long and fondly remembered by connoisseurs of judicial delicatessen—that is, as the performances of Weber and Fields are remembered by students of dramatic science.* In immediate retrospect, it grows more fantastic and exhilarating. Scopes has had precisely the same fair trial that the Hon. John Philip Hill, accused of bootlegging on the oath of Howard A. Kelly, would have before the Rev. Dr. George W. Crabbe. He is a fellow not without humor; I find him full of smiles today. On some near tomorrow the Sheriff will collect a month's wages from him, but he has certainly had a lot of fun.

More interesting than the hollow buffoonery that remains will be the effect upon the people of Tennessee, the actual prisoners at the bar. That the more civilized of them are in a highly feverish condition of mind must be patent to every visitor. The guffaws that roll in from all sides give them great pain. They are full of bitter protests and valiant projects. They prepare, it appears, to organize, hoist the black flag and offer the fundamentalists of the dung-hills a battle to the death. They will not cease until the last Baptist preacher is in flight over the mountains, and the ordinary intellectual decencies of Christendom are triumphantly restored.

* Popular comedy duo, Lew Fields and Joe Weber.

With the best will in the world I find it impossible to accept this tall talk with anything resembling confidence. The intelligentsia of Tennessee had their chance and let it get away from them. When the old mountebank, Bryan, first invaded the State with his balderdash they were unanimously silent. When he began to round up converts in the back country they offered him no challenge. When the Legislature passed the anti-evolution bill and the Governor signed it, they contented themselves with murmuring *pianissimo*. And when the battle was joined at last and the time came for rough stuff only one Tennesseean of any consequence volunteered.

That lone volunteer was Dr. John Neal, now of counsel for the defense, a good lawyer and an honest man. His services to Darrow, Malone and Hays have been very valuable and they come out of the case with high respect for him. But how does Tennessee regard him? My impression is that Tennessee vastly underestimates him. I hear trivial and absurd criticism of him on all sides and scarcely a word of praise for his courage and public spirit. The test of the State is to be found in its attitude toward such men. It will come out of the night of Fundamentalism when they are properly appreciated and honored, and not before. When that time comes I'll begin to believe that the educated minority here is genuinely ashamed of the Bryan obscenity and that it is prepared to combat other such disgraces hereafter resolutely in the open and regardless of the bellowing of the mob.

The Scopes trial, from the start, has been carried on in a manner exactly fitted to the anti-evolution law and the simian imbecility under it. There hasn't been the slightest pretense to decorum. The rustic judge, a candidate for re-election, has postured before the yokels like a clown in a ten-cent side show, and almost every word he has uttered has been an undisguised appeal to their prejudices and superstitions. The chief prosecuting attorney, beginning like a competent lawyer and a man of self-respect, ended like a convert at a Billy Sunday revival. It fell to him, finally, to make a clear and astounding statement of theory of justice prevailing under Fundamentalism. What he said, in brief, was that a man accused of infidelity had no rights whatever under Tennessee law.

This probably not true yet, but it will become true inevitably if the Bryan murrain is not arrested. The Bryan of today is not to be mistaken for the political rabble-rouser of two decades ago. That earlier Bryan may have been grossly in error, but he at least kept his errors within the bounds of reason: it was still possible to follow him without yielding up all intelligence. The Bryan of today, old, disappointed and embittered, is a far different bird. He realizes at last the glories of this world are not for him, and he takes refuge, peasant-like, in religious hallucinations. They depart from sense altogether. They are not merely silly; they are downright idiotic. And, being idiotic, they appeal with irresistible force to the poor half-wits upon whom the old charlatan now preys. When I heard him, in open court, denounce the notion that man is a mammal I was genuinely staggered and so was every other stranger in the courtroom. People looked at one another in blank amazement. But the native Fundamentalists, it quickly appeared, saw nothing absurd in his words. The attorneys for the prosecution smiled approval, the crowd applauded, the very judge on the bench beamed his acquiescence. And the same thing happened when he denounced all education as corrupting and began arguing incredibly that a farmer who read the Bible knew more than any scientist in the world. Such dreadful bilge, heard of far away, may seem only ridiculous. But it takes on a different smack, I assure you, when one hears it discharged formally in a court of law and sees it accepted as wisdom by judge and jury.

Darrow has lost this case. It was lost long before he came to Dayton. But it seems to me that he has nevertheless performed a great public service by fighting it to a finish and in a perfectly serious way. Let no one mistake it for comedy, farcical though it may be in all its details.

It serves notice on the country that Neanderthal man is organizing in these forlorn backwaters of the land, led by a fanatic, rid sense and devoid of conscience. Tennessee, challenging him too timorously and too late, now sees its courts converted into camp meetings and its Bill of Rights made a mock of by its sworn officers of the law. There are other States that had better look to their arsenals before the Hun is at their gates.

XIII

Tennessee in
the Frying Pan

From *The Baltimore Evening Sun*, July 20, 1925

I

That the rising town of Dayton, when it put the infidel Scopes on trial, bit off far more than it has been able to chew—this melancholy fact must now be evident to everyone. The village Aristides Sophocles Goldsboroughs believed that the trial would bring in a lot of money, and produce a vast mass of free and profitable advertising. They were wrong on both counts, as boomers usually are. Very little money was actually spent by the visitors: the adjacent yokels brought their own lunches and went home to sleep, and the city men from afar rushed down to Chattanooga whenever there was a lull. As for the advertising that went out over the leased wires, I greatly fear that it has quite ruined the town. When people recall it hereafter they will think of it as they think of Herrin, Ill., and Homestead, Pa. It will be a joke town at best, and infamous at worst.*

The natives reacted to this advertising very badly. The preliminary publicity, I believe, had somehow disarmed and deceived them. It was mainly amiable spoofing; they took it philosophically, assured by the local Aristideses that it was good for trade. But when the main guard of Eastern and Northern journalists swarmed down, and their dispatches began to show the country and the world exactly how the obscene buffoonery appeared to realistic city men, then the yokels began to sweat coldly, and in a few days they were full of terror and indignation. Some of the bolder spirits, indeed, talked gaudily of direct action against the authors of the "libels." But the history of the Ku Klux and the American Legion offers overwhelmingly evidence that 100 per cent Americans never fight when the enemy is in strength, and able to make a defense, so the visitors suffered nothing worse than black, black looks. When the last of them departs Daytonians will disinfect the town with sulphur candles, and the local pastors will exorcise the devils that they left behind them.

---

*Herrin was the site of a 1922 massacre of strikebreakers by union sympathizers. Homestead was the site of an 1892 battle between Pinkertons hired by Andrew Carnegie and steelworkers.

II

Dayton, of course, is only a ninth-rate country town, and so its agonies are of relatively little interest to the world. Its pastors, I daresay, will be able to console it, and if they fail there is always the old mountebank, Bryan, to give a hand. Faith cannot only move mountains; it can also soothe the distressed spirits of mountaineers. The Daytonians, unshaken by Darrow's ribaldries, still believe. They believe that they are not mammals. They believe, on Bryan's word, that they know more than all the men of science of Christendom. They believe, on the authority of Genesis, that the earth is flat and that witches still infest it. They believe, finally and especially, that all who doubt these great facts of revelation will go to hell. So they are consoled.

But what of the rest of the people of Tennessee? I greatly fear that they will not attain to consolation so easily. They are an extremely agreeable folk, and many of them are highly intelligent. I met men and women— particularly women—in Chattanooga who showed every sign of the highest culture. They led civilized lives, despite Prohibition, and they were interested in civilized ideas, despite the fog of Fundamentalism in which they moved. I met members of the State judiciary who were as heartily ashamed of the bucolic ass, Raulston, as an Osler would be of a chiropractor. I add the educated clergy: Episcopalians, Unitarians, Jews and so on—enlightened men, tossing pathetically under the imbecilities of their evangelical colleagues. Chattanooga, as I found it, was charming, but immensely unhappy.

What its people ask for—many of them in plain terms—is suspended judgment, sympathy, Christian charity, and I believe that they deserve all these things. Dayton may be typical of Tennessee, but it is surely not all of Tennessee. The civilized minority in the State is probably as large as in any other Southern State. What ails it is simply the fact it has been, in the past, too cautious and politic—that it has been too reluctant to offend the Fundamentalist majority. To that reluctance something else has been added: an uncritical and somewhat childish local patriotism. The Tennesseeans have tolerated their imbeciles for fear that attacking them would bring down the derision of the rest of the country. Now they have the derision, and to excess— and the attack is ten times as difficult as it ever was before.

III

How they are to fight their way out of their wallow I do not know. They begin the battle with the enemy in command of every height and every gun; worse, there is a great deal of irresolution in their own ranks. The newspapers of the State, with few exceptions, are very feeble. One of the best of them, the *Chattanooga News*, set up an eloquent whooping for Bryan the moment he got to Dayton. Before that it had been against the anti-evolution law. But with the actual battle joined, it began to wobble, and presently it was printing articles arguing that Fundamentalism, after all, made men happy—that a Tennesseean gained something valuable by being an ignoramus—in other words, that a hog in a barnyard was to be envied by an Aristotle. The *News* was far better than most: it gave space, too, to the other side, and at considerable risk. But its weight, for two weeks, was thrown heavily to Bryan and his balderdash.

The pusillanimous attitude of the bar of the State I described in my dispatches from Dayton. It was not until the trial was two days old that any Tennessee lawyers of influence and dignity went to the aid of Dr. John R. Neal— and even then all of the volunteers enlisted only on condition that their names be kept out of the newspapers. I should except one T.B. McElwee. He sat at the trial table and rendered valuable services. The rest lurked in the background. It was an astounding situation to a Marylander, but it seemed to be regarded as quite natural in Tennessee.

The prevailing attitude toward Neal himself was also very amazing. He is an able lawyer and a man of repute, and in any Northern State his courage would get the praise it deserves. But in Tennessee even the intelligentsia seem to feet that he has done something discreditable by sitting at the trial table with Darrow, Hays and Malone. The State buzzes with trivial, idiotic gossip about him— that he dresses shabbily, that he has political aspirations, and so on. What if he does and has? He has carried himself, in this case, in a way that does higher credit to his native State. But his native State, instead of being proud of him, simply snarls at him behind his back.

IV

So with every other man concerned with the defense— most of them, alackaday, foreigners. For example, Rappelyea, the Dayton engineer who was first to go to the aid of Scopes. I was told solemnly in Dayton, not once but twenty times, that Rappelyea was (a) a Bowery boy from New York, and (b) an incompetent and ignorant engineer. I went to some trouble to unearth the facts. They were (a) that he was actually a member of one of the oldest Huguenot families in America, and (b) that his professional skill and general culture were such that the visiting scientists sought him out and found pleasure in his company.

Such is the punishment that falls upon a civilized man cast among fundamentalists. As I have said, the worst of it is that even the native intelligentsia help to pull the rope. In consequence all the brighter young men of the State— and it produces plenty of them—tend to leave it. If they remain, they must be prepared to succumb to the prevailing blather or resign themselves to being more or less infamous. With the anti-evolution law enforced, the State university will rapidly go to pot; no intelligent youth will waste his time upon its courses if he can help it. And so, with the young men lost, the struggle against darkness will become almost hopeless.

As I have said, the State still produces plenty of likely young bucks—if only it could hold them! There is good blood everywhere, even in the mountains. During the dreadful buffooneries of Bryan and Raulston last week two typical specimens sat at the press table. One was Paul Y. Anderson, correspondent of the *St. Louis Post-Dispatch*, and the other was Joseph Wood Krutch, one of the editors of *The Nation*. I am very familiar with the work of both of them, and it is my professional judgment that it is of the first caliber. Anderson is one of the best newspaper reporters in America and Krutch is one of the best editorial writers.

Well, both were there as foreigners. Both were working for papers that could not exist in Tennessee. Both were viewed by their fellow Tennesseeans not with pride, as credits to the State, but as traitors to the Tennessee *Kultur* and public enemies. Their crime was that they were intelligent men, doing their jobs intelligently.

XIV

Bryan

From *The Baltimore Evening Sun*, July 27, 1925

I

It was plain to everyone, when Bryan came to Dayton, that his great days were behind him—that he was now definitely an old man, and headed at last for silence. There was a vague, unpleasant manginess about his appearance; he somehow seemed dirty, though a close glance showed him carefully shaved, and clad in immaculate linen. All the hair was gone from the dome of his head, and it had begun to fall out, too, behind his ears, like that of the late Samuel Gompers. The old resonance had departed from his voice: what was once a bugle blast had become reedy and quavering. Who knows that, like Demosthenes, he had a lisp? In his prime, under the magic of his eloquence, no one noticed it. But when he spoke at Dayton it was always audible.

When I first encountered him, on the sidewalk in front of the Hicks brothers' law office, the trial was yet to begin, and so he was still expansive and amiable. I had printed in *The Nation,* a week or so before, an article arguing that the anti-evolution law, whatever its unwisdom, was at least constitutional—that policing school teachers was certainly not putting down free speech. The old boy professed to be delighted with the argument, and gave the gaping bystanders to understand that I was a talented publicist. In turn I admired the curious shirt he wore—sleeveless and with the neck cut very low. We parted in the manner of two Spanish ambassadors.

But that was the last touch of affability that I was destined to see in Bryan. The next day the battle joined and his face became hard. By the end of the first week he was simply a walking malignancy. Hour by hour he grew more bitter. What the Christian Scientists call malicious animal magnetism seemed to radiate from him like heat from a stove. From my place in the court-room, standing upon a table, I looked directly down upon him, sweating horribly and pumping his palm-leaf fan. His eyes fascinated me: I watched them all day long. They were blazing points of hatred. They glittered like occult and sinister gems. Now and then they wandered to me, and I got my share. It was like coming under fire.

II

What was behind that consuming hatred? At first I thought that it was mere evangelical passion. Evangelical Christianity, as everyone knows, is founded upon hate, as the Christianity of Christ was founded upon love. But even evangelical Christians occasionally loose their belts and belch amicably; I have known some who, off duty, were very benignant. In that very courtroom, indeed, were some of them—for example, old Ben McKenzie, Nestor of the Dayton bar, who sat beside Bryan. Ben was full of good humor. He made jokes with Darrow. But Bryan only glared.

One day it dawned on me that Bryan, after all, was an evangelical Christian only by sort of afterthought—that his career in this world, and the glories thereof, had actually come to an end before he ever began whooping for Genesis. So I came to this conclusion: that what really moved him was a lust for revenge. The men of the cities had destroyed him and made a mock of him; now he would lead the yokels against them. Various facts clicked into the theory, and I hold it still. The hatred in the old man's burning eyes was not for the enemies of God; it was for the enemies of Bryan.

Thus he fought his last fight, eager only for blood. It quickly became frenzied and preposterous, and after that pathetic. All sense departed from him. He bit right and left, like a dog with rabies. He descended to demagogy so dreadful that his very associates blushed. His one yearning was to keep his yokels heated up—to lead his forlorn mob against the foe. That foe, alas, refused to be alarmed. It insisted upon seeing the battle as a comedy. Even Darrow, who knew better, occasionally yielded to the prevailing spirit. Finally, he lured poor Bryan into a folly almost incredible.

I allude to his astounding argument against the notion that man is a mammal. I am glad I heard it, for otherwise I'd never believe it. There stood the man who had been thrice a candidate for the Presidency of the Republic—and once, I believe, elected—there he stood in the glare of the world, uttering stuff that a boy of eight would laugh at! The artful Darrow led him on: he repeated it, ranted for it, bellowed it in his cracked voice. A tragedy, indeed! He came into life a hero, a Galahad, in bright and shining armor. Now he was passing out a pathetic fool.

III

Worse, I believe that he somehow sensed the fact—that he realized his personal failure, whatever the success of the grotesque cause he spoke for. I had left Dayton before Darrow's cross-examination brought him to his final absurdity, but I heard his long speech against the admission of expert testimony, and I saw how it fell flat and how Bryan himself was conscious of the fact. When he sat down he was done for, and he knew it. The old magic had failed to work; there was applause but there was no exultant shouts. When, half an hour later, Dudley Field Malone delivered his terrific philippic, the very yokels gave him five times the clapper-clawing that they had given to Bryan.

This combat was the old leader's last, and it symbolized in more than one way his passing. Two women sat through it, the one old and crippled, the other young and in the full flush of beauty. The first was Mrs. Bryan; the second was Mrs. Malone. When Malone finished his speech the crowd stormed his wife with felicitations, and she glowed as only a woman can who has seen her man fight a hard fight and win gloriously. But no one congratulated Mrs. Bryan. She sat hunched in her chair near the judge, apparently very uneasy. I thought then that she was ill—she has been making the round of sanitariums for years, and was lately in the hands of a faith-healer—but now I think that some appalling prescience was upon her, and that she saw in Bryan's eyes a hint of the collapse that was so near.

He sank into his seat a wreck, and was presently forgotten in the blast of Malone's titanic rhetoric. His speech had been maundering, feeble and often downright idiotic. Presumably, he was speaking to a point of law, but it was quickly apparent that he knew no more law than the bailiff at the door. So he launched into mere violet garrulity. He dragged in snatches of ancient chautauqua addresses; he wandered up hill and down dale. Finally, Darrow lured him into that fabulous imbecility about man as a mammal. He sat down one of the most tragic asses in American history.

IV

It is the national custom to sentimentalize the dead, as it is to sentimentalize men about to be hanged. Perhaps I fall into that weakness here. The Bryan I shall remember is the Bryan of his last weeks on earth—broken, furious, and infinitely pathetic. It was impossible to meet his hatred with hatred to match it. He was winning a battle that would make him forever infamous wherever enlightened men remembered it and him. Even his old enemy, Darrow, was gentle with him at the end. That cross-examination might have been ten times as devastating. It was plain to everyone that the old Berseker Bryan was gone—that all that remained of him was a pair of glaring and horrible eyes.

But what of his life? Did he accomplish any useful thing? Was he, in his day, of any dignity as a man, and of any value to his fellow-men? I doubt it. Bryan, at his best, was simply a magnificent job-seeker. The issues that he bawled about usually meant nothing to him. He was ready to abandon them whenever he could make votes by doing so, and to take up new ones at a moment's notice. For years he evaded Prohibition as dangerous; then he embraced it as profitable. At the Democratic National Convention last year he was on both sides, and distrusted by both. In his last great battle there was only a baleful and ridiculous malignancy. If he was pathetic, he was also disgusting.

Bryan was a vulgar and common man, a cad undiluted. He was ignorant, bigoted, self-seeking, blatant and dishonest. His career brought him into contact with the first men of his time; he preferred the company of rustic ignoramuses. It was hard to believe, watching him at Dayton, that he had traveled, that he had been received in civilized societies, that he had been a high officer of state. He seemed only a poor clod like those around him, deluded by a childish theology, full of an almost pathological hatred of all learning, all human dignity, all beauty, all fine and noble things. He was a peasant come home to the dung-pile. Imagine a gentleman, and you have imagined everything that he was not.

The job before democracy is to get rid of such canaille. If it fails, they will devour it.

XV

Round Two

From *The Baltimore Evening Sun*, August 10, 1925

I

The translation of Bryan to a higher sphere was a body blow to the imbecility called Fundamentalism, and its effects are already visible. Not only has the Georgia Legislature incontinently rejected the anti-evolution bill; there has been a marked improvement in the discussion of the whole subject throughout the South. While Bryan lived it was almost impossible, in most Southern States, to make any headway against him. His great talent for inflaming the mob, and his habit of doing it by lying about his opponents, made many Southern editors hesitate to tackle him. In a region where education is backward, and popular thinking is largely colored by disreputable politicians and evangelical pastors, such a fellow was dangerous.

But a dead man cannot bite, and so the Southern editors now show a new boldness. I speak, of course, of the general. A few daring spirits have been denouncing Bryan as a charlatan for a long while, and some of them have even carried their readers with them. I point, for example, to Julian Harris in Columbus, Ga,, and to Charlton Wright in Columbia, S.C.—two highly civilized men, preaching sense and decency without fear. But the average Southern editor, it must be manifest, has been, in the past, of a different sort. What ails the South, primarily, is simply lack of courage. Its truculence is only protective coloration; it is really very timid. If there had been bolder editors in Tennessee there would have been no anti-evolution bill and no Scopes trial.

But, as I say, the removal of Bryan to Paradise gives heart to skittish spirits, for his heirs and assigns are all palpable fifth-raters, and hence not formidable. In South Carolina, for example, the cause falls to the Hon. Cole L. Blease, who is to Bryan what a wart is to the Great Smokey Mountains. In Tennessee itself he is succeeded by a junta of hedge lawyers, county school superintendents, snide politicians and rustic clergymen—in brief, by worms. It will be easy to make practice against them.

II

The circumstances of Bryant's death, indeed, have probably done great damage to Fundamentalism, for it is nothing if it is not a superstition, and the rustic pastors will have a hard time explaining to the faithful why the agent of God was struck down in the midst of the first battle. How is it that Darrow escaped and Bryan fell? There is, no doubt, a sound theological reason, but I shouldn't like to have to expound it, even to a country Bible class. In the end, perhaps, the true believers will have to take refuge from the torment of doubt in the theory that the hero was murdered, say by the Jesuits. Even so, there will be the obvious and disquieting inference that, in the first battle, the devil really won.

The theory I mention is already launched. I find it in the current issue of the *American Standard*, a leading fundamentalist organ, edited by an eminent Baptist pastor. This journal, which is written in good English and attractively printed, voices the opinion of the more refined and thoughtful Fundamentalists. What it says today is said by scores of little denominational papers tomorrow. Its notion is that the Catholics, represented by Dudley Field Malone, and the Jews, represented by Darrow (!), concentrated such malicious animal magnetism upon poor Bryan that he withered and perished. The late martyr Harding, it appears, was disposed of in the same way: his crime was that he was a Freemason. Thus Fundamentalism borrows the magic of Christian Science, and idiot kisses idiot.

But something remains for the rev. clergy to explain, and that is Bryan's vulnerability. If he was actually divinely inspired, and doing battle for the True Faith, then how come that he did not throw off Malone's and Darrow's sorceries? He had ample warning. Dayton, during the Scopes trial, was full of whispers. At least a dozen times I was told of hellish conspiracies afoot. Every pastor in the town knew that demons filled the air. Why didn't they exorcise these dreadful shapes? One must assume that they prayed for the champion of light. In fact, they prayed openly, and in loud, ringing, confident tones. Then why did their prayers fail?

III

I do not propound such questions in an effort to be jocose; I offer them as characteristic specimens of Fundamentalist reasoning. The Fundamentalist prayer is not an inner experience; it is a means to objective ends. he prays precisely as more worldly Puritans complain to the police. he expects action, and is disappointed and dismayed if it does not follow. The mind of the Fundamentalist is extremely literal—indeed, the most literal mind ever encountered on this earth. He doubts nothing in the Bible, not even typographical errors. He believes absolutely that Noah took two behemoths and two streptococci into the Ark, and he believes with equal faith that the righteous have angels to guard them.

Thus the dramatic death of Bryan is bound to give him great concern, and in the long run, I believe, it will do more to break down his cocksureness than ten thousand arguments. Try to imagine the debates that must be going on in Dayton itself, in Robinson's drug store and on the courthouse lawn. What is old Ben McKenzie's theory? How does the learned Judge Raulston, J., explain it? And the Hicks boys? And Pastor Stribling? I venture to guess that the miracle—for everything that happens, to a Fundamentalist, is a miracle—has materially cooled off enthusiasm for the Bryan Fundamentalist University. If Darrow could blast Bryan, then what is to prevent him blasting the university, and so setting fire to town?

My belief is that the last will soon be heard of that great institution. It will engage the newspapers for a few weeks or months longer, and various enterprising souls will get a lot of free publicity by subscribing to its endowment, and then it will be quietly shelved. I doubt that anyone in Tennessee wants it—that is, anyone who has any notion what a university is.

The yokels of the hills may be bemused by it, as they are bemused by the scheme to put God into the Constitution. But the rest of the Tennesseeans are eager only to shove Fundamentalism into a cellar, and to get rid of the disgrace that it has brought upon the State.

IV

They will tackle it with more vigor than last time when it next takes to the warpath. They have learned a lesson. Already, indeed, they make plans for the repeal of the anti-evolution law—a long business, but certainly not hopeless. It was supported by the politicians of the State simply because those in favor of it were noisy and determined, and those against it were too proud to fight. These politicians will begin to wobble the moment it becomes clear that there are two sides engaged, and they will desert Genesis at the first sign that the enemy has artillery, and is eager to use it. They are, like politicians everywhere, men without conscience. One of the chief of them began life as a legislative agent for the brewers. When Prohibition came in he became a violent Prohibitionist.

Their brethren elsewhere in the South are of the same sort; it is hard to find, in that whole region, a politician who is an honest man. the news from Georgia shows which way the wind is blowing. If it had seemed to them that Fundamentalism was prospering, the Georgia legislators would have rammed through the anti-evolution bill with a whoop. But the whisper reached them that there were breakers ahead, and so they hesitated, and the measure was lost. Those breakers were thrown up by a few determined men, notably the Julian Harris aforesaid, son of Joel Chandler Harris.* What he accomplished in Georgia, almost single-handedly, will not be lost upon the civilized minorities of the other Southern States. Imbecility has raged down there simply because no one has challenged it. Challenged, it will have hard going, there as elsewhere.

With Bryan alive and on the warpath, inflaming the morons and spreading his eloquent nonsense, the battle would have been ten times harder. But Bryan was unique, and can have no successor. His baleful rhetoric died with him; in fact, it died a week before his corporeal frame. In a very true sense Darrow killed him. When he emerged from that incredible cross-examination, all that was most dangerous in his old following deserted him. It was no longer possible for a man of any intelligence to view him as anything save a pathetic has-been.

---

* Joel Chandler Harris (1845-1908) was a popular author of children's tales told in dialect.

XVI

Aftermath

From *The Baltimore Evening Sun*, September 14, 1925

I

The Liberals, in their continuing discussion of the late trial of the infidel Scopes at Dayton, Tenn., run true to form. That is to say, they show all their habitual lack of humor and all their customary furtive weakness for the delusions of *Homo neanderthalensis*. I point to two of their most enlightened organs: the eminent *New York World* and the gifted *New Republic*. The *World* is displeased with Mr. Darrow because, in his appalling cross-examination of the mountebank Bryan, he did some violence to the theological superstitions that millions of Americans cherish. The *New Republic* denounces him because he addressed himself, not to "the people of Tennessee" but to the whole country, and because he should have permitted "local lawyers" to assume "the most conspicuous position in the trial."

Once more, alas, I find myself unable to follow the best Liberal thought. What the *World*'s contention amounts to, at bottom, is simply the doctrine that a man engaged in combat with superstition should be very polite to superstition. This, I fear, is nonsense. The way to deal with superstition is not to be polite to it, but to tackle it with all arms, and so rout it, cripple it, and make it forever infamous and ridiculous. Is it, perchance, cherished by persons who should know better? Then their folly should be brought out into the light of day, and exhibited there in all its hideousness until they flee from it, hiding their heads in shame.

True enough, even a superstitious man has certain inalienable rights. He has a right to harbor and indulge his imbecilities as long as he pleases, provided only he does not try to inflict them upon other men by force. He has a right to argue for them as eloquently as he can, in season and out of season. He has a right to teach them to his children. But certainly he has no right to be protected against the free criticism of those who do not hold them. He has no right to demand that they be treated as sacred. He has no right to preach them without challenge. Did Darrow, in the course of his dreadful bombardment of Bryan, drop a few shells, incidentally, into measurably cleaner camps? Then let the garrisons of those camps look to their defenses. They are free to shoot back. But they can't disarm their enemy.

II

The meaning of religious freedom, I fear, is sometimes greatly misapprehended. It is taken to be a sort of immunity, not merely from governmental control but also from public opinion. A dunderhead gets himself a long-tailed coat, rises behind the sacred desk, and emits such bilge as would gag a Hottentot. Is it to pass unchallenged? If so, then what we have is not religious freedom at all, but the most intolerable and outrageous variety of religious despotism. Any fool, once he is admitted to holy orders, becomes infallible. Any half-wit, by the simple device of ascribing his delusions to revelation, takes on an authority that is denied to all the rest of us.

I do not know how many Americans entertain the ideas defended so ineptly by poor Bryan, but probably the number is very large. They are preached once a week in at least a hundred thousand rural churches, and they are heard too in the meaner quarters of the great cities. Nevertheless, though they are thus held to be sound by millions, these ideas remain mere rubbish. Not only are they not supported by the known facts; they are in direct contravention of the known facts. No man whose information is sound and whose mind functions normally can conceivably credit them. They are the products of ignorance and stupidity, either or both.

What should be a civilized man's attitude toward such superstitions? It seems to me that the only attitude possible to him is one of contempt. If he admits that they have any intellectual dignity whatever, he admits that he himself has none. If he pretends to a respect for those who believe in them, he pretends falsely, and sinks almost to their level. When he is challenged he must answer honestly, regardless of tender feelings. That is what Darrow did at Dayton, and the issue plainly justified the act. Bryan went there in a hero's shining armor, bent deliberately upon a gross crime against sense. He came out a wrecked and preposterous charlatan, his tail between his legs. Few Americans have ever done so much for their country in a whole lifetime as Darrow did in two hours.

III

The caveat of the *New Republic* is so absurd that it scarcely deserves an answer. It is based upon a complete misunderstanding of the situation that the Scopes trial revealed. What good would it have done to have addressed an appeal to the people of Tennessee? They had already, by their lawful representatives, adopted the anti-evolution statute by an immense majority, and they were plainly determined to uphold it. The newspapers of the State, with one or two exceptions, were violently in favor of the prosecution, and applauded every effort of the rustic judge and district attorney to deprive the defense of its most elemental rights.

True enough, there was a minority of Tennesseeans on the other side—men and women who felt keenly the disgrace of their State, and were eager to put an end to it. But their time had passed; they had missed their chance. They should have stepped forward at the very beginning, long before Darrow got into the case. Instead, they hung back timorously, and so Bryan and the Baptist pastors ran amok. There was a brilliant exception: John R. Neal. There was another: T.R. Elwell. Both lawyers. But the rest of the lawyers of the State, when the issue was joined at last, actually helped the prosecution. Their bar associations kept up a continuous fusillade. They tried their best to prod the backwoods Dogberry, Raulston, into putting Darrow into jail.

There was but one way to meet this situation and Darrow adopted it. He appealed directly to the country and to the world. He had at these recreant Tennesseeans by exhibiting their shame to all men, near and far. He showed them cringing before the rustic theologians, and afraid of Bryan. He turned the State inside out, and showed what civilization can come to under Fundamentalism.

The effects of that cruel exposure are now visible. Tennessee is still spluttering—and blushing. The uproar staggered its people. And they are doing some very painful thinking. Will they cling to Fundamentalism or will they restore civilization? I suspect that the quick decision of their neighbor, Georgia, will help them to choose. Darrow did more for them, in two weeks, than all their pastors and politicians had done since the Civil War.

IV

His conduct of the case, in fact, was adept and intelligent from beginning to end. It is hard, in retrospect, to imagine him improving it. He faced immense technical difficulties. In order to get out of the clutches of the village Dogberry and before judges of greater intelligence he had to work deliberately for the conviction of his client. In order to evade the puerile question of that client's guilt or innocence and so bring the underlying issues before the country, he had to set up a sham battle on the side lines. And in order to expose the gross ignorance and superstition of the real prosecutor, Bryan, he had to lure the old imposter upon the stand.

It seems to me that he accomplished all of these things with great skill. Scopes was duly convicted, and the constitutional questions involved in the law will now be heard by competent judges and decided without resort to prayer and moving pictures. The whole world has been made familiar with the issues, and the nature of the menace that Fundamentalism offers to civilization is now familiar to every schoolboy. And Bryan was duly scotched, and, if he had lived, would be standing before the country today as a comic figure, tattered and preposterous.

All this was accomplished, in infernal weather, by a man of sixty-eight, with the scars of battles all over him. He had, to be sure, highly competent help. At his table sat lawyers whose peculiar talents, in combination, were of the highest potency—the brilliant Hays, the eloquent Malone, the daring and patriotic Tennesseean, Neal. But it was Darrow who carried the main burden, and Darrow who shaped the final result. When he confronted Bryan at last, the whole combat came to its climax. On the one side was bigotry, ignorance, hatred, superstition, every sort of blackness that the human mind is capable of. On the other side was sense. And sense achieved a great victory.

XVII

To Expose a Fool

From *The American Mercury*, October, 1925

I

Has it been marked by historians that the late William
Jennings Bryan's last secular act on this earth was to
catch flies? A curious detail, and not without its sardonic
overtones. He was the most sedulous flycatcher in
American history, and by long odds the most successful.
His quarry, or course, was not *Musca domestica* but *Homo
neandertalensis*. For forty years he tracked it with snare and
blunderbuss, up and down the backways of the Republic.
Wherever the flambeaux of Chautauqua smoked and gut-
tered, and the bilge of Idealism ran in the veins, and
Baptist pastors dammed the brooks with the saved, and

men gathered who were weary and heavy laden, and their wives who were unyieldingly multiparous and full of Peruna—there the indefatigable Jennings set up his traps and spread his bait. He knew every forlorn country town in the South and West, and he could crowd the most remote of them to suffocation by simply winding his horn. The city proletariat, transiently flustered by him in 1896, quickly penetrated his buncombe and would have no more of him; the gallery jeered at him at every Democratic National Convention for twenty-five years. But out where the grass grows high, and the horned cattle dream away the lazy day, and men still fear the powers and principles of the air—out there between the corn-rows he held his old puissance to the end. There was no need of beaters to drive his game. The news that he was coming was enough. For miles the flivver dust would choke the roads. And when he rose at the end of the day to discharge his Message there would be such a breathless attention, such a rapt and enchanted ecstasy, such a sweet rustle of amens as the world has not known since Johannan fell to Herod's headsman.

There was something peculiarly fitting in the fact that his last days were spent in a one-horse Tennessee village, and that death found him there. The man felt at home in such scenes. He liked people who sweated freely, and were not debauched by the refinements of the toilet. Making his progress up and down the Main Street of little Dayton, surrounded by gaping primates from the upland valleys of the Cumberland Range, his coat laid aside, his bare arms and hairy chest shining damply, his bald head sprinkled with dust—so accoutred and on display he was obviously happy. He liked getting up early in the morning, to the tune of cocks crowing on the dunghill. He liked the heavy, greasy victuals of the farmhouse kitchen. He liked country lawyers, country pastors, all country people. I believe that this liking was sincere—perhaps the only sincere thing in the man. His nose showed no uneasiness when a hillman in faded overalls and hickory shirt accosted him on the street, and besought him for light upon some mystery of Holy Writ. The simian gabble of a country town was not gabble to him, but wisdom of an occult and superior sort. In the presence of city folks he was palpably uneasy. Their clothes, I suspect, annoyed him, and he was suspicious of their too delicate manners. He knew all the while that they were laughing at him—if not at his baroque theology, then at least at his alpaca pantaloons. But the yokels never laughed at him. To them he was not the huntsman but the prophet, and toward the end, as he gradually forsook mundane politics for purely ghostly concerns, they began to elevate him in their hierarchy. When he died he was the peer of Abraham. Another curious detail: his old enemy, Wilson,

aspiring to the same white and shining robe, came down
with a thump. But Bryan made the grade. His place in the
Tennessee hagiocracy is secure. If the village barber saved
any of his hair, then it is curing gall-stones down there
today.

II

But what label will he bear in more urbane regions? One,
I fear, of a far less flattering kind. Bryan lived too long,
and descended too deeply into the mud, to be taken seri-
ously hereafter by fully literate men, even of the kind
who write school-books. There was a scattering of sweet
words in his funeral notices, but it was not more than a
response to conventional sentimentality. The best verdict
the most romantic editorial writer could dredge up, save
in the eloquent South, was to the general effect that his
imbecilities were excused by his earnestness—that under
his clowning, as under that of the juggler of Notre Dame,
there was the zeal of a steadfast soul. But this was apology, not praise; precisely the same thing might be said of
Mary Baker G. Eddy, the late Czar Nicholas, or Czolgoz.
The truth is that even Bryan's sincerity will probably
yield to what is called, in other fields, definitive
criticism. Was he sincere when he opposed imperialism
in the Philippines, or when he fed it with deserving
Democrats in Santo Domingo? Was he sincere when he

tried to shove the Prohibitionists under the table, or
when he seized their banner and began to lead them with
loud whoops? Was he sincere when he bellowed against
war, or when he dreamed himself into a tin-soldier in
uniform, with a grave reserved among the generals? Was
he sincere when he denounced the late John W. Davis, or
when he swallowed Davis? Was he sincere when he fawned
over Champ Clark, or when he betrayed Clark? Was he
sincere when he pleaded for tolerance in New York, or
when he bawled for the fagot and the stake in Tennessee?

This talk of sincerity, I confess, fatigues me. If the fellow
was sincere, then so was P.T. Barnum. The word is disgraced
and degraded by such uses. He was, in fact, a charlatan, a
mountebank, a zany without any shame or dignity. What
animated him from end to end of his grotesque career was
simply ambition—the ambition of a common man to get
his hand upon the collar of his superiors, or, failing that, to
get his thumb into their eyes. He was born with a roaring
voice, and it had the trick of inflaming half-wits against
their betters, that he himself might shine. His last battle
will be grossly misunderstood if it is thought of as a
mere exercise in fanaticism—that is, if Bryan the
Fundamentalist Pope is mistaken for one of the bucolic
Fundamentalists. There was much more in it than that, as
everyone knows who saw him on the field. What moved
him, at bottom, was simply hatred of city men who had

laughed at him so long, and brought him at last to so tatterdemalion an estate. He lusted for revenge upon them. He yearned to lead the anthropoid rabble against them, to set *Homo neandertalensis* upon them, to punish them for the execution they had done upon him by attacking the very vitals of their civilization. He went far beyond the bounds of any merely religious frenzy, however inordinate. When he began denouncing the notion that man is a mammal even some of the hinds at Dayton were agape. And when, brought upon Darrow's cruel hook, he writhed and tossed in a very fury of malignancy, bawling against the baldest elements of sense and decency like a man frantic—when he came to the tragic climax there were snickers among the hinds as well as hosannas.

Upon that hook, in truth, Byran committed suicide, as a legend as well as in the body. He staggered from the rustic court ready to die, and he staggered from it ready to be forgotten, save as a character in a third-rate farce, witless and in execrable taste. The chances are that history will put the peak of democracy in his time; it has been on the downward curve among us since the campaign of 1896. He will be remembered, perhaps, as its supreme impostor, the *reductio ad adsurdum* of its pretension. Bryan came very near being President of the United States. In 1896, it is possible, he was actually elected. He lived long enough to make patriots thank the inscrutable gods for Harding, even for Coolidge. Dullness has got into the White House, and the smell of cabbage boiling, but there is at least nothing to compare to the intolerable buffoonery that went on in Tennessee. The President of the United States doesn't believe that the earth is square, and that witches should be put to death, and that Jonah swallowed the whale. The Golden Text is not painted weekly on the White House wall, and there is no need to keep ambassadors waiting while Pastor Simpson, of Smithville, prays for rain in the Blue Room. We have escaped something—by a narrow margin, but still safely

III

That is, so far. The Fundamentalists continue at the wake, and sense gets a sort of reprieve. The legislature of Georgia, so the news comes, has shelved the anti-evolution bill, and turns its back upon the legislature of Tennessee. Elsewhere minorities prepare for battle—here and there with some assurance of success. But it is too early, it seems to me, to send the firemen home; the fire is still burning on many a far-flung hill, and it may begin to roar again at any moment. The evil that men do lives after them. Bryan, in his malice, started something that will not be easy to stop. In ten thousand country towns his old heelers, the evangelical pastors, are propagating his gospel, and everywhere the yokels are ready for it. When he disappeared from the big cities, the big cities made the capital error of assuming that he was done for. If they heard of him at all, it was only as a crimp for real-estate speculators— the heroic foe of the unearned increment hauling it in with both hands. He seemed preposterous, and hence harmless. But all the while he was busy among his old lieges, preparing for a *jacquerie* that should floor all his enemies at one blow. He did the job competently. He had vast skill at such enterprises. Heave an egg out of a Pullman window, and you will hit a Fundamentalist almost anywhere in the United States today. They swarm in the country towns, inflamed by their pastors, and with a saint, now, to venerate. They are thick in the mean streets behind the gas-works. They are everywhere that learning is too heavy a burden for mortal minds, even the vague, pathetic learning on tap in little red schoolhouses. They march with the Klan, with the Christian Endeavor Society, with the Junior Order of United American Mechanics, with the Epworth League, with all the rococo bands that poor and unhappy folk organize to bring some light of purpose into their lives.* They have had a thrill, and they are ready for more.

---

* The Epworth League was a Methodist youth organization focused on religious development.

Such is Bryan's legacy to his country. He couldn't be President, but he could at least help magnificently in the solemn business of shutting off the presidency from every intelligent and self respecting man. The storm, perhaps, won't last long, as times goes in history. It may help, indeed, to break up the democratic delusion, now already showing weakness, and so hasten its own end. But while it lasts it will blow off some roofs and flood some sanctuaries.

ABOVE LEFT: William Jennings Bryan arriving
in Dayton on July 7, 1925. He was greeted
at the train station by town leaders, a band
playing religious and patriotic songs, and a
crowd of over 1,000 people. (BRYAN COLLEGE)

ABOVE RIGHT: Dayton, Tennessee. Robinson's
Drug Store is on the right. Owner Fred Robinson
was one of many who thought the trial would
bring a boom in tourism to the town, and he
tried to capitalize on it by stringing an adver-
tisement over Main Street. (BRYAN COLLEGE)

ABOVE LEFT: John Scopes, left, walks to the courthouse with one of his attorneys, John Neal, center, and George Rappelyea, right.
The "Read Your Bible" banner is typical of the many trial-related posters displayed around Dayton during the trial, and identical to one that hung in the courtroom until the judge begrudgingly removed it upon the strenuous objection of the defense. (BRYAN COLLEGE)

ABOVE CENTER: Henry Louis Mencken
(ENOCH PRATT FREE LIBRARY)

ABOVE RIGHT: Friday, July 10, 1925:
The jury is sworn in for the case of
Tennessee vs. John Thomas Scopes.
(BRYAN COLLEGE)

ABOVE: Bryan addresses the bench during the trial. Prosecutor Ben McKenzie sits with arms folded to Bryan's left.

ABOVE RIGHT: Clarence Darrow addresses the court. Note the standing room only crowd. The trial was held in the Rhea Country Courthouse in Tennessee's largest courtroom, which had seats for over 500 people – but no ceiling fans.

RIGHT: One of the tensest moments in the trial occurred when Darrow was threatened with a contempt charge for saying he didn't think his client was able to get a fair trial in Raulston's courtroom. Here, standing from left to right, defense attorney Dudley Field Malone and prosecutors J. Gordon McKenzie, Wallace Haggard, Herbert Hicks, and Tom Stewart listen to

Darrow's forced apology to the court. (Note the WGN microphone. The Scopes trial was the first ever broadcast live to a national audience. The man below the microphone stand is commonly misidentified as Mencken.)

FAR RIGHT: Judge John T. Raulston issues a ruling from the bench. Seen here with the two policeman that he kept stationed on either side of himself with fans. Raulston carried a Bible into the courtroom every day, brought his family to listen to a Bryan sermon attacking the defense at the local Methodist church (sitting in the front row), and began the trial each day with a prayer—despite the vociferous objection of the defense.

(ALL PHOTOS: BRYAN COLLEGE)

ABOVE LEFT: Temperatures throughout the
trial were in the 100-degree range. Here, Bryan
shows one of the effects of the heat.

ABOVE RIGHT: Due to the heat in the courtroom and to accommodate growing crowds, on
July 20, 1925, Raulston moved the trial
outdoors for what were expected to be closing
arguments. Darrow surprised the court, however,
by calling Bryan to the stand. Despite objections
by his fellow prosecutors, Bryan agreed to
testify. The two-hour exchange would go
down in history, but not in the court record—
Raulston expunged it the next day, saying it
was irrelevant. Darrow immediately urged the
jury to find his client guilty so that, as planned
all along, he could appeal to a higher court.

ABOVE LEFT: July 21, 1925: Scopes listens to the jury render its verdict of guilty, which it did after only nine minutes of deliberation. Raulston issued a fine of $100, the minimum. After the verdict, Scopes addressed the court for the first and only time during the trial, saying, "Your honor, I feel that I have been convicted of violating an unjust statute. I will continue in the future, as I have in the past, to oppose this law in any way I can. Any other action would be in violation of my ideal of academic freedom—that is, to teach the truth as guaranteed in our constitution of personal and religious freedom." The ruling was overturned two years later. The state never collected the fine. (BRYAN COLLEGE)

ABOVE RIGHT: Five days after the trial ended, Bryan died in his sleep during a Sunday afternoon nap. He had remained in Dayton to prepare a pamphlet of his closing argument, prefatory to a national speaking tour planned to capitalize on trial publicity. Here, on the same platform where a band had greeting him less than three weeks before, his casket is loaded onto a special train taking him to Washington, D.C., for burial in Arlington National Cemetery. Thousands lined the route to bid him farewell. (BRYAN COLLEGE)

Appendix

The Examination of William Jennings Bryan by Clarence Darrow

Monday, July 20, 1925

JUDGE RAULSTON: Do you want Mr. Bryan sworn?

DARROW: No.

BRYAN: I can make affirmation; I can say, "So help me God, I will tell the truth."

DARROW: No, I take it you will tell the truth. You have given considerable study to the Bible, haven't you, Mr. Bryan?

BRYAN: Yes, sir, I have tried to.

DARROW: Well, we all know you have; we are not going to dispute that at all. But you have written and published articles almost weekly, and sometimes have made interpretations of various things.

BRYAN: I would not say interpretations, Mr. Darrow, but comments on the lesson.

DARROW: If you comment to any extent, those comments have been interpretations?

BRYAN: I presume that my discussion might be to some extent interpretations, but they have not been primarily intended as interpretations.

DARROW: But you have studied that question, of course?

BRYAN: Of what?

DARROW: Interpretation of the Bible.

BRYAN: On this particular question?

DARROW: Yes, sir.

BRYAN: Yes, sir.

DARROW: Then you have made a general study of it?

BRYAN: Yes, I have. I have studied the Bible for about fifty years, or some time more than that. But, of course, I have studied it more as I have become older than when I was but a boy.

DARROW: Do you claim that everything in the Bible should be literally interpreted?

BRYAN: I believe everything in the Bible should be accepted as it is given there. Some of the Bible is given illustratively; for instance, "Ye are the salt of the earth." I would not insist that man was actually salt, or that he had flesh of salt, but it is used in the sense of salt as saving God's people.

DARROW: But when you read that Jonah swallowed the

whale—or that the whale swallowed Jonah, excuse me, please—how do you literally interpret that?

BRYAN: When I read that a big fish swallowed Jonah—it does not say whale.

DARROW: Doesn't it? Are you sure?

BRYAN: That is my recollection of it, a big fish. And I believe it, and I believe in a God who can make a whale and can make a man, and can make both do what He pleases.

DARROW: Mr. Bryan, doesn't the New Testament say whale?

BRYAN: I am not sure. My impression is that it says fish, but it does not make so much difference. I merely called your attention to where it says fish, it does not say whale.

DARROW: But in the New Testament it says whale, doesn't it?

BRYAN: That may be true. I cannot remember in my own mind what I read about it.

DARROW: Now, you say the big fish swallowed Jonah, and he remained how long—three days—and then he spewed him up on the land. You believe that the big fish was made to swallow Jonah?

BRYAN: I am not prepared to say that; the Bible merely says it was done.

DARROW: You don't know whether it was the ordinary run of fish or made for that purpose?

BRYAN: You may guess; you evolutionists guess.

DARROW: But when we do guess, we have the sense to guess right.

BRYAN: But you do not do it often.

DARROW: You are not prepared to say whether that fish was made especially to swallow a man or not?

BRYAN: The Bible doesn't say, so I am not prepared to say.

DARROW: You don't know whether that was fixed up specially for the purpose.

BRYAN: No, the Bible doesn't say.

DARROW: But do you believe He made them—that He made such a fish, and that it was big enough to swallow Jonah?

BRYAN: Yes, sir. And let me add, one miracle is just as easy to believe as another.

DARROW: It is for me.

BRYAN: It is for me, too.

DARROW: Just as hard?

BRYAN: It is hard to believe for you, but easy for me. A miracle is a thing performed beyond what man can perform. When you get beyond what man can do, you get within the realms of miracles; and it is just as easy to believe the miracle of Jonah as any other miracle in the Bible.

DARROW: Perfectly easy to believe that Jonah swallowed the whale?

BRYAN: The Bible says so. The Bible doesn't make as extreme statements as evolutionists do.

DARROW: That may be a question, Mr. Bryan, about some of those you have known.

BRYAN: The only thing is, you have a definition of fact that includes imagination.

DARROW: And you have a definition that excludes everything but imagination!

STEWART (FOR THE PROSECUTION): I object to that as argumentative.

DARROW: The witness must not argue with me, either. Do you consider the story of Jonah and the whale a miracle?

BRYAN: I think it is.

DARROW: Do you believe Joshua made the sun stand still?

BRYAN: I believe what the Bible says. I suppose you mean that the earth stood still?

DARROW: I don't know. I'm talking about the Bible now.

BRYAN: I accept the Bible absolutely.

DARROW: The Bible says Joshua commanded the sun to stand still for the purpose of lengthening the day, doesn't it, and you believe it?

BRYAN: I do.

DARROW: Do you believe at that time the entire sun went around the earth?

BRYAN: No, I believe that the earth goes around the sun.

DARROW: Do you believe that the men who wrote it thought that the day could be lengthened or that the sun could be stopped?

BRYAN: I don't know what they thought.

DARROW: You don't know?

BRYAN: I think they wrote the fact without expressing their own thoughts.

DARROW: Have you an opinion as to whether or not the men who wrote that thought...

STEWART: I want to object, Your Honor. It has gone beyond the pale of any issue that could possibly be injected into this lawsuit, except by imagination. I do not think the defendant has a right to conduct the examination any further, and I ask Your Honor to exclude it.

JUDGE RAULSTON: I will hear Mr. Bryan.

BRYAN: It seems to me it would be too exacting to confine the defense to the facts. If they are not allowed to get away from the facts, what have they to deal with?

JUDGE RAULSTON: Mr. Bryan is willing to be examined. Go ahead.

DARROW: Have you an opinion as to whether whoever wrote the book, I believe it was Joshua—the Book of Joshua—thought the sun went around the earth or not?

BRYAN: I believe that he was inspired.

DARROW: Can you answer my question?

BRYAN: When you let me finish the statement.

DARROW: It is a simple question, but finish it.

BRYAN: You cannot measure the length of my answer by the length of your question. [Laughter.]

DARROW: No, except that the answer will be longer. [Laughter.]

BRYAN: I believe that the Bible is inspired, and an inspired author, whether one who wrote as he was directed to write, understood the things he was writing about, I don't know.

DARROW: Whoever inspired it, do you think whoever inspired it believed that the sun went around the earth?

BRYAN: I believe it was inspired by the Almighty, and he may have used language that could be understood at that time, instead of using language that could not be understood until Darrow was born. [Laughter and applause.]

DARROW: So it might not—it might be subject to construction, might it not?

BRYAN: It might have been used in language that could be understood then.

DARROW: That means it is subject to construction?

BRYAN: That is your construction. I am answering your question.

DARROW: Is that correct?

BRYAN: That is my answer to it.

DARROW: Can you answer?

BRYAN: I might say Isaiah spoke of God sitting upon the circle of the earth.

DARROW: I am not talking about Isaiah.

JUDGE RAULSTON: Let him illustrate if he wants to.

DARROW: It is your opinion that the passage was subject to construction?

BRYAN: Well, I think anybody can put his own construction upon it, but I do not mean that necessarily it is a correct construction. I have answered the question.

DARROW: Don't you believe that in order to lengthen the day, it would have been construed that the earth stood still?

BRYAN: I would not attempt to say what would have been necessary, but I know this: that I can take a glass of water that would fall to the ground without the strength of my hand, and to the extent of the glass of water I can overcome the law of gravitation and lift it up, whereas without my hand, it would fall to the ground. If my puny hand can overcome the law of gravitation, the most universally understood, to that extent, I would not set a limit to the power of the hand of the Almighty God, that made the universe.

DARROW: I read that years ago, in your "Prince of Peace." Can you answer my question directly? If the day was lengthened by stopping either the earth or the sun, it must have been the earth?

BRYAN: Well, I should say so. Yes, but it was language that was understood at that time, and we now know that the sun stood still, as it was, with the earth.

DARROW: We know also the sun does not stand still.

BRYAN: Well, it is relatively so, as Mr. Einstein would say.

DARROW: I ask you if it does stand still?

BRYAN: You know as well as I know.

DARROW: Better. You have no doubt about it?

BRYAN: No, no.

DARROW: And the earth moves around it?

BRYAN: Yes, but I think there is nothing improper if you will protect the Lord against against your criticism.

DARROW: I suppose He needs it?

BRYAN: He was using language at that time that the people understood.

DARROW: And that you call "interpretation?"

BRYAN: No, sir, I would not call it interpretation.

DARROW: I say you would call it interpretation at this time, to say it meant something then?

BRYAN: You may use your own language to describe what I have to say, and I will use mine in answering.

DARROW: Now, Mr. Bryan, have you ever pondered what would have happened to the earth if it had stood still?

BRYAN: No.

DARROW: You have not?

BRYAN: No, sir. The God I believe in could have taken care of that, Mr. Darrow.

DARROW: I see. Have you ever pondered what would naturally happen to the earth if it stood still suddenly?

BRYAN: No.

DARROW: Don't you know it would have been converted into a molten mass of matter?

BRYAN: You testify to that when you get on the stand; I will give you a chance.

DARROW: Don't you believe it?

BRYAN: I would want to hear expert testimony on that.

DARROW: You have never investigated that subject?

BRYAN: I don't think I have ever had the question asked.

DARROW: Or ever thought of it?

BRYAN: I have been too busy on things that I thought were of more importance than that.

DARROW: You believe the story of the flood to be a literal interpretation?

BRYAN: Yes, sir.

DARROW: When was that flood?

BRYAN: I wouldn't attempt to fix the date. The date is fixed, as suggested this morning.

DARROW: About 2400 B.C.?

BRYAN: That has been the estimate of a man that is accepted today. I would not say it is accurate.

DARROW: That estimate is printed in the Bible?

BRYAN: Everybody knows. At least I think most of the people know that was the estimate given.

DARROW: But what do you think that the Bible itself says? Do you know how that estimate was arrived at?

BRYAN: I never made a calculation.

DARROW: A calculation from what?

BRYAN: I could not say.

DARROW: From the generations of man?

BRYAN: I would not want to say that.

DARROW: What do you think?

BRYAN: I do not think about things I don't think about.

DARROW: Do you think about things you do think about?

BRYAN: Well, sometimes. [Laughter.]

DARROW: Mr. Bryan, you have read these dates over and over again?

BRYAN: Not very accurately. I turn back sometimes to see what the time was.

DARROW: You want to say now, you have no idea how these dates were computed?

BRYAN: No, I don't say. But I have told you what my idea was. I say I don't know how accurate it was.

DARROW: You say from the generation of man....

STEWART: I am objecting to his cross-examining his own witnesses.

DARROW: He is a hostile witness.

JUDGE RAULSTON: I am going to let Mr. Bryan control.

BRYAN: I want to give him all the latitude that he wants, for I am going to have some latitude when he gets through.

DARROW: You can have latitude and longitude. [Laughter.]

JUDGE RAULSTON: Order.

STEWART: The witness is entitled to be examined as to the legal evidence of it. We were supposed to go into the origin of this case, and we have nearly lost the day, Your Honor.

McKENZIE (FOR THE PROSECUTION): I object to it.

STEWART: Your Honor, he is perfectly able to take care of this, but we are attaining no evidence. This is not competent evidence.

BRYAN: These gentlemen have not had much chance. They did not come here to try this case. They came here to try revealed religion. I am here to defend it, and they can ask me any questions they please.

JUDGE RAULSTON: All right. [Applause.]

DARROW: Great applause from the bleachers.

BRYAN: From those whom you call "yokels."

DARROW: I have never called them yokels.

BRYAN: That is the ignorance of Tennessee, the bigotry.

DARROW: You mean who are applauding you?

BRYAN: Those are the people whom you insult.

DARROW: You insult every man of science and learning in the world because he does not believe in your fool religion.

JUDGE RAULSTON: I will not stand for that.

DARROW: For what he is doing?

JUDGE RAULSTON: I am talking to both of you.

STEWART: This has gone beyond the pale of a lawsuit, your Honor. I have a public duty to perform under my oath, and I ask the court to stop it. Mr. Darrow is making an effort to insult the gentleman on the witness stand, and I ask that this be stopped, for it has gone beyond the pale of a lawsuit.

JUDGE RAULSTON: To stop it now would not be just to Mr. Bryan. He wants to ask the other gentleman questions along the same line.

STEWART: It will all be incompetent.

BRYAN: The jury is not here.

JUDGE RAULSTON: I do not want to be strictly technical.

DARROW: Then Your Honor rules and I accept. How long ago was the flood, Mr. Bryan?

BRYAN: Let me see Ussher's calculation about it.*

DARROW: Surely. [Hands a Bible to Bryan.]

* James Ussher (1581-1656) was an Irish bishop and scholar; his calculations of biblical dates were incorporated into an authorized version of the Bible.

BRYAN: I think this does not give it.

DARROW: It gives an account of Noah. Where is the one in evidence? I am quite certain it is there.

BRYAN: Oh, I would put the estimate where it is, because I have no reason to vary it. But I would have to look at it to give you the exact date.

DARROW: I would, too. Do you remember what book the account is in?

BRYAN: Genesis... It is given here as 2348 years before Christ.

DARROW: Well, 2348 years BC. You believe that all the living things that were not contained in the ark were destroyed?

BRYAN: I think the fish may have lived.

DARROW: Outside of the fish?

BRYAN: I cannot say.

DARROW: You cannot say?

BRYAN: No, except that just as it is, I have no proof to the contrary.

DARROW: I am asking you whether you believe it.

BRYAN: I do.

DARROW: That all living things outside of the fish were destroyed.

BRYAN: What I say about the fish is merely a matter of humor.

DARROW: I understand.

BRYAN: Due to the fact that a man wrote up here the other day to ask whether all the fish were destroyed, and the gentleman who received the letter told him the fish may have lived.

DARROW: I am referring to the fish too.

BRYAN: I accept that as the Bible gives it, and I have never found any reason for denying, disputing, or rejecting it.

DARROW: Let us make it definite: 2,348 years?

BRYAN: I didn't say that. That is the time given, but I don't pretend to say that is exact.

DARROW: You never figured it out, those generations, by yourself?

BRYAN: No, sir, not myself.

DARROW: But the Bible you have offered in evidence says 2340 something, so that 4200 years ago there was not a living thing on earth, excepting the people on the ark and the animals on the ark, and the fishes.

BRYAN: There had been living things before that.

DARROW: I mean at that time.

BRYAN: After that.

DARROW: Don't you know there are any number of civilizations that are traced back to more than 5,000 years?

BRYAN: I know we have people who trace things back according to the number of ciphers they have. But I am not satisfied they are accurate.

DARROW: You are not satisfied that there is any civilization that can be traced back five thousand years?

BRYAN: I would not want to say there is, because I have no evidence of it that is satisfactory.

DARROW: Would you say there is not?

BRYAN: Well, so far as I know, but when the scientists differ from twenty-four millions to three hundred millions in their opinions as to how long ago life came here, I want them to be nearer, to come nearer together, before they demand of me to give up my belief in the Bible.

DARROW: Do you say that you do not believe that there were any civilizations on this earth that reach back beyond five thousand years?

BRYAN: I am not satisfied by any evidence that I have seen.

DARROW: I didn't ask you what you are satisfied with— I asked you if you believed it.

BRYAN: Will you let me answer it?

JUDGE RAULSTON: Go right on.

BRYAN: I am satisfied by no evidence that I have found that would justify me in accepting the opinions of these men against what I believe to be the inspired word of God.

DARROW: And you believe every nation, every organization of men, every animal in the world outside of the fishes—

BRYAN: The fish, I want you to understand, is merely a matter of humor.

DARROW: I believe the Bible says so. Take the fishes in?

BRYAN: Let us get together and look over this.

DARROW: Probably we would better. We will after we get through. You believe that all the various human races on the earth have come into being in the last four thousand years or four thousand two hundred years, whatever it is?

BRYAN: No; it would be more than that. Sometime after the creation, before the flood.

DARROW: 1925 added to it?

BRYAN: The flood is 2300 and something; and creation, according to the estimate there, is further back than that.

DARROW: Then you don't understand me. If we don't get together on it, look at the book. This is the year of grace 1925, isn't it? Let us put down 1925. Have you got a pencil? [One of the defense attorneys hands Darrow a pencil.]

BRYAN: Add that to 4,004?

DARROW: Yes.

BRYAN: That is the date given here on the first page, according to Bishop Ussher, which I say I accept only because I have no reason to doubt it.

DARROW: 1925 plus 4004 is 5,929 years. Now then, what do you subtract from that?

BRYAN: That is the beginning.

DARROW: I was talking about the flood.

BRYAN: 2348 on that, we said.

DARROW: Less that?

BRYAN: No, subtract that from 4000. It would be about 1700 years.

DARROW: That is the same thing.

BRYAN: No. Subtracted, it is 2300 and something before the beginning of the Christian era, about 1700 years after the Creation.

DARROW: If I add 2300 years, that is the beginning of the Christian era?

BRYAN: Yes, sir.

DARROW: If I add 1925 to that I will get it, won't I?

BRYAN: Yes, sir.

DARROW: That makes 4,262 years?

BRYAN: According to the Bible there was a civilization before that, destroyed by the flood.

DARROW: Let me make this definite. You believe that every civilization on the earth and every living thing, except possibly the fishes, that came out of the ark, were wiped out by the flood?

BRYAN: At that time.

DARROW: At that time; and then whatever human beings, including all the tribes that inhabited the world, and have inhabited the world, and who run their pedigree straight back, and all the animals, have come on to the earth since the flood?

BRYAN: Yes.

DARROW: Within 4200 years. Do you know a scientific man on the earth that believes any such thing?

BRYAN: I cannot say. But I know some scientific men who dispute entirely the antiquity of man as testified to by other scientific men.

DARROW: Only that does not answer the question. Do you know of a single scientific man on the face of the earth that believes any such thing as you stated, about the antiquity of man?

BRYAN: I don't think I have ever asked one the direct question.

DARROW: Quite important, isn't it?

BRYAN: Well, I don't know as it is.

DARROW: It might not be?

BRYAN: If I had nothing else to do except speculate on what our remote ancestors were and what our remote descendants have been, but I have been more interested in Christians going on right now, to make it much more important than speculations on either the past or the future.

DARROW: You have never had any interest in the age of the various races and people and civilizations and animals that exist upon the earth today. Is that right?

BRYAN: I have never felt a great deal of interest in the effort that has been made to dispute the Bible by the speculations of men, or the investigations of men.

DARROW: Are you the only human being on earth who knows what the Bible means?

STEWART: I object.

JUDGE RAULSTON: Sustained.

DARROW: You do know that there are thousands of people who profess to be Christians who believe the earth is much more ancient and that the human race is much more ancient?

BRYAN: I think there may be.

DARROW: And you never have investigated to find out how long man has been on the earth?

BRYAN: I have never found it necessary to examine every speculation; but if I had done it I never would have done anything else.

DARROW: I ask for a direct answer.

BRYAN: I do not expect to find out all those things. I do not expect to find out about races.

DARROW: I didn't ask you that. Now, I ask you if you know, if it was interesting enough, or important enough for you to try to find out how old these ancient civilizations are?

BRYAN: No, I have not made a study of it.

DARROW: Don't you know that the ancient civilizations of China are six or seven thousand years old at the very least?

BRYAN: No; but they would not run back beyond the creation, according to the Bible six thousand years.

DARROW: You don't know how old they are, is that right?

BRYAN: I don't know how old they are, but possibly you do. [Laughter.] I think you would give the preference to anybody who opposed the Bible, and I give the preference to the Bible.

DARROW: I see. Well, you are welcome to your opinion. Have you any idea how old the Egyptian civilization is?

BRYAN: No.

DARROW: Do you know of any record in the world, outside of the story of the Bible, which conforms to any statement that it is 4,200 years ago or thereabouts, that all life was wiped off the face of the earth?

BRYAN: I think they have found records.

DARROW: Do you know of any?

BRYAN: Records reciting the flood, but I am not an authority on the subject.

DARROW: Now, Mr. Bryan: will you say if you know of any record, or have ever heard of any records that describe that a flood existed 4,200 years ago, or about that time, which wiped all life off the earth?

BRYAN: The recollection of what I have read on the subject is not distinct enough to say whether the records attempted to fix a time, but I have seen in the discoveries of archaeologists where they have found records that described the flood.

DARROW: Mr. Bryan, don't you know that there are many old religions that describe the flood?

BRYAN: No, I don't know.

DARROW: You know there are others besides the Jewish?

BRYAN: I don't know whether those are the record of any other religion, or refer to this flood.

DARROW: Don't you ever examine religion so far to know that?

BRYAN: Outside of the Bible?

DARROW: Yes.

BRYAN: No, I have not examined to know that, generally.

DARROW: You have never examined any other religions?

BRYAN: Yes, sir.

DARROW: Have you ever read anything about the origins of religions?

BRYAN: Not a great deal.

DARROW: You have never examined any other religion?

BRYAN: Yes, sir.

DARROW: And you don't know whether any other religion gave a similar account of the destruction of the earth by the flood?

BRYAN: The Christian religion has satisfied me and I have never felt it necessary to look up some competing religions.

DARROW: Do you consider that every religion on earth competes with the Christian religion?

BRYAN: I think everybody who believes in the Christian religion believes so...

DARROW: I am asking what you think.

BRYAN: I do not regard them as competitive because I do not think they have the same source as we have.

DARROW: You are wrong in saying "competitive"?

BRYAN: I would not say competitive, but the religious unbelievers.

DARROW: Unbelievers of what?

BRYAN: In the Christian religion.

DARROW: What about the religion of Buddha?

BRYAN: Well, I can tell you something about that, if you would like to know.

DARROW: What about the religion of Confucius or Buddha?

BRYAN: Well, I can tell you something about them, if you would like to know.

DARROW: Did you ever investigate them?

BRYAN: Somewhat.

DARROW: Do you regard them as competitive?

BRYAN: No, I think they are very inferior. Would you like for me to tell you what I know about it?

DARROW: No.

BRYAN: Well, I shall insist on giving it to you.

DARROW: You won't talk about free silver, will you?

BRYAN: Not at all.

STEWART: I object to counsel going any further and cross-examining his own witness. He is your own witness.

DARROW: Well, now, general, I assume that every lawyer knows perfectly well that we have a right to cross-examine a hostile witness. Is there any doubt about that?

STEWART: Under the law in Tennessee, if you put a witness on and he proves to be hostile to you, the law provides the method by which you may cross-examine him. You will have to make an affidavit that you are surprised at his statement, and you may do that.

BRYAN: Is there any way by which a witness can make an affidavit that the attorney also is hostile?

DARROW: I am not hostile to you. I am hostile to your views, and I suppose that runs with me, too.

BRYAN: But I think when the gentleman asked me about Confucius I ought to be allowed to answer his question.

DARROW: Oh, tell it, Mr. Bryan, I won't object to it.

BRYAN: I had occasion to study Confucianism when I went to China. I got all I could find about what Confucius said, and I found that there were several direct and strong contrasts between the teachings of

Jesus and the teachings of Confucius. In the first place, one of his followers asked if there was any word that would express all that was necessary to know in the relations of life, and he said, "Isn't reciprocity such a word?" I know of no better illustration of the difference between Christianity and Confucianism than the contrast that is brought out there. Reciprocity is a calculating selfishness. If a person does something for you, you do something for him and keep it even. That is the basis for the philosophy of Confucius. Christ's doctrine was not of reciprocity. We were told to help people not in proportion as they had helped us—not in proportion as they might have helped us, but in proportion to their needs, and there is all the difference in the world between a religion that teaches you just to keep even with other people and the religion that teaches you to spend yourself for other people and to help them as they need help.

DARROW: There is no doubt about that. I haven't asked you that.

BRYAN: That is one of the differences between the two.

DARROW: Do you know how old the Confucian religion is?

BRYAN: I can't give you the exact date of it.

DARROW: Did you ever investigate to find out?

BRYAN: Not to be able to speak definitely as to date, but I can tell you something I read, and will tell you.

DARROW: Wouldn't you just as soon answer my questions, and get along?

BRYAN: Yes, sir.

DARROW: Of course, if I take any advantage of misquoting you, I don't object to being stopped. Do you know how old the religion of Zoroaster is?

BRYAN: No, sir.

DARROW: Do you know they are both more ancient than the Christian religion?

BRYAN: I am not willing to take the opinion of people who are trying to find excuses for rejecting the Christian religion, when they attempt to give dates and hours and minutes. And they will have to get together and be more exact than they have yet been able, to compel me to accept just what they say as if it were absolutely true.

DARROW: Are you familiar with James Clark's book on the ten great religions?

BRYAN: No.

DARROW: He was a Unitarian minister, wasn't he? You don't think he was trying to find fault, do you?

BRYAN: I am not speaking of the motives of men.

DARROW: You don't know how old they are, all these other religions?

BRYAN: I wouldn't attempt to speak correctly, but I think it is much more important to know the difference between them than to know the age.

DARROW: Not for the purpose of this inquiry, Mr. Bryan. Do you know about how many people there were on this earth at the beginning of the Christian era?

BRYAN: No, I don't think I ever saw a census on that subject.

DARROW: Do you know how many people there were on this earth 3,000 years ago?

BRYAN: No.

DARROW: Did you ever try to find out?

BRYAN: When you display my ignorance, could you not give me the facts so I would not be ignorant any longer? Can you tell me how many people there were when Christ was born?

DARROW: You know, some of us might get the facts and still be ignorant.

BRYAN: Will you please give me that? You ought not to ask me a question that you don't know the answer to.

DARROW: I can make an estimate.

BRYAN: What is your estimate?

DARROW: Wait until you get to me. Do you know anything about how many people there were in Egypt 3500 years ago, or how many people there were in China 5000 years ago?

BRYAN: No.

DARROW: Have you ever tried to find out?

BRYAN: No, sir, you are the first man I ever heard of who was interested in it. [Laughter.]

DARROW: Mr. Bryan, am I the first man you ever heard of who has been interested in the age of human societies and primitive man?

BRYAN: You are the first man I ever heard speak of the number of people at these different periods.

DARROW: Where have you lived all your life?

BRYAN: Not near you. [Laughter.]

DARROW: Nor near anybody of learning?

BRYAN: Oh, don't assume you know it all.

DARROW: Do you know that there are thousands of books in your libraries on all these subjects I have been asking you about?

BRYAN: I couldn't say, but I will take your word for it.

DARROW: Did you ever read a book on primitive man? Like Tylor's "Primitive Culture" or Boas or any of the great authorities?

BRYAN: I don't think I have ever read the ones you have mentioned.

DARROW: Have you read any?

BRYAN: Well, I have read a little from time to time, but I didn't pursue it, because I didn't know I was to be called as a witness.

DARROW: You have never in all your life made any attempt to find out about the other peoples of the earth—how old their civilizations are, how long they have existed on the earth, have you?

BRYAN: No, sir, I have been so well satisfied with the Christian religion that I have spent no time trying to find arguments against it.

DARROW: Were you afraid you might find some?

BRYAN: No, sir, I am not afraid now that you will show me any.

DARROW: You remember that man who said—I am not quoting literally—that one could not be content though he rose from the dead? You suppose you could be content?

BRYAN: Well, will you give the rest of it, Mr. Darrow?

DARROW: No.

BRYAN: Why not?

DARROW: I am not interested.

BRYAN: Why scrap the Bible? They have Moses and the Prophets.

DARROW: Who has?

BRYAN: That is the rest of the quotation you didn't finish.

DARROW: And so you think if they have Moses and the Prophets, they don't need to find out anything else?

BRYAN: That was the answer that was made there.

DARROW: And you followed the same vein?

BRYAN: I have all the information I want to live by and to die by.

DARROW: And that's all you are interested in?

BRYAN: I am not looking for any more on religion.

DARROW: You don't care how old the earth is, how old man is, or how long the animals have been here?

BRYAN: I am not so much interested in that.

DARROW: You have never made any investigation to find out?

BRYAN: No, sir, I have never.

DARROW: All right.

BRYAN: Now, will you let me finish the question?

DARROW: What question was that? If there is anything more you want to say about Confucius, I don't object.

BRYAN: Oh yes, I have got two more things.

DARROW: If Your Honor please, I don't object, but his speeches are not germane to my question.

HICKS (FOR THE PROSECUTION): Your Honor, he put him on.

JUDGE RAULSTON: You went into it and I will let him explain.

DARROW: I asked him certain specific questions about Confucius.

HICKS: The questions he is asking are not germane either.

DARROW: I think they are.

BRYAN: I mentioned the word "reciprocity" to show the

difference between Christ's teaching in that respect and the teachings of Confucius. I call your attention to another difference. One of the followers of Confucius asked him, "What do you think of the doctrine that you should reward evil with good?" And the answer of Confucius was, "Reward evil with justice and reward good with good. Love your enemies. Overcome evil with good. And there is a difference between the two teachings—a difference incalculable in its effect and in—the third difference—people who scoff at religion and try to make it appear that Jesus brought nothing into the world, talk about the Golden Rule of Confucius. Confucius said, "Do not unto others what you would not have others do unto you." There is all the difference in the world between a negative harmlessness and a positive helpfulness, and the Christian religion is a religion of helpfulness, of service, embodied in the language of Jesus when he said, "Let him who would be chiefest among you be the servant of all." Those are the three differences between the teachings of Jesus and the teaching of Confucius, and they are very strong differences on very important questions. Now, Mr. Darrow, you asked me if I knew anything about Buddha.

DARROW: You want to make a speech on Buddha, too?

BRYAN: No sir, I want to answer your question on Buddha.

DARROW: I asked you if you knew anything about him.

BRYAN: I do.

DARROW: Well, that's answered, then.

BRYAN: Buddha...

DARROW: Well, wait a minute. You answered the question.

RAULSTON: I will let him tell what he knows.

DARROW: All he knows?

RAULSTON: Well, I don't know about that.

BRYAN: I won't insist on telling all I know. I will tell more than Mr. Darrow wants told.

DARROW: Well, all right, tell it. I don't care.

BRYAN: Buddhism is an agnostic religion.

DARROW: To what? What do you mean by "agnostic"?

BRYAN: I don't know.

DARROW: You don't know what you mean?

BRYAN: That is what "agnosticism" is—"I don't know." When I was in Rangoon, Burma, one of the Buddhists told me that they were going to send a delegation to an agnostic congress that was to be held soon at Rome and I read in an official document...

DARROW: Do you remember his name?

BRYAN: No sir, I don't.

DARROW: What did he look like? How tall was he?

BRYAN: I think he was about as tall as you, but not so good-looking.

DARROW: Do you know about how old a man he was?

Do you know whether he was old enough to know what he was talking about?

BRYAN: He seemed to be old enough to know what he was talking about. [Laughter.]

DARROW: If Your Honor please, instead of answering plain specific questions we are permitting the witness to regale the crowd with what some black man said to him when he was travelling in Rangoon, India.

BRYAN: He was dark-colored, but not black.

JUDGE RAULSTON: I will let him go ahead and answer.

BRYAN: I wanted to say that I then read a paper that he gave me, and official paper of the Buddhist church, and it advocated the sending of delegates to that agnostic conference at Rome, arguing that it was an agnostic religion and I will give you another evidence of it. I went to call on a Buddhist teacher.

DARROW: I object to Mr. Bryan making a speech every time I ask him a question.

JUDGE RAULSTON: Let him finish his answer and then you can go ahead.

BRYAN: I went to call on a Buddhist priest and found him at his noon meal, and there was an Englishman there who was also a Buddhist. He went over as ship's carpenter and became a Buddhist and had been for about six years, and while I waited for the Buddhist priest I talked to the Englishman and he said the most important thing was you didn't have to believe to be a Buddhist.

DARROW: You know the name of the Englishman?

BRYAN: No sir, I don't know his name.

DARROW: What did he look like? What did he look like?

BRYAN: He was what I would call an average looking man.

DARROW: How could you tell he was an Englishman?

BRYAN: He told me so.

DARROW: Do you know whether he was truthful or not?

BRYAN: No sir, but I took his word for it.

JUDGE RAULSTON: Well, get along, Mr. Darrow, with your examination.

DARROW: Mr. Bryan ought to get along. You have heard of the Tower of Babel, haven't you?

BRYAN: Yes, sir.

DARROW: That tower was built under the ambition that they could build a tower up to heaven, wasn't it? And God saw what they were at, and to prevent their getting into heaven He confused their tongues?

BRYAN: Something like that. I wouldn't say to prevent their getting into heaven. I don't think it is necessary to believe that God was afraid they would get to heaven.

DARROW: I mean that way?

BRYAN: I think it was a rebuke to them.

DARROW: A rebuke to them trying to go that way?

BRYAN: To build the tower for that purpose.

DARROW: To take that short cut?

BRYAN: That is your language, not mine.

DARROW: Now, when was that?

BRYAN: Give us the Bible.

DARROW: Yes, we will have strict authority on it. Scientific authority?

BRYAN: That was about 100 years before the flood, Mr. Darrow, according to this chronology. It was 2247—the date on one page is 2218 and on the other, 2247. And it is described in here—

DARROW: That is the year 2247?

BRYAN: 2218 BC is at the top of one page and 2247 at the other, and there is nothing in here to indicate the change.

DARROW: Well, make it 2230 then?

BRYAN: All right, about.

DARROW: Then you add 1500 to that.

BRYAN: No, 1925.

DARROW: Add 1925 to that, that would be 4155 years ago. Up to 4155 years ago every human being on earth spoke the same language?

BRYAN: Yes, sir, I think that is the inference that could be drawn from that.

DARROW: All the different languages of the earth, dating from the Tower of Babel, is that right? Do you know how many languages are spoken on the face of the earth?

BRYAN: No. I know the Bible has been translated into 500 and no other book has been translated into anything like that many.

DARROW: That is interesting, if true. Do you know all the languages there are?

BRYAN: No, sir, I can't tell you. There may be many dialects besides that in some languages, but those are all the principal languages.

DARROW: There are a great many that are not principal languages?

BRYAN: Yes sir.

DARROW: You haven't any idea how many there are?

BRYAN: No, sir.

DARROW: How many people have spoken those various languages?

BRYAN: No, sir.

DARROW: And you say that all those languages of all the sons of men have come on the earth not over 4150 years ago?

BRYAN: I have seen no evidence that would lead me to put it any farther back than that.

DARROW: That is your belief, anyway—that was due to the confusion of tongues at the Tower of Babel. Did you ever study philology at all?

BRYAN: No, I have never made a study of it; not in the sense in which you speak of it.

DARROW: You have used language all your life?

BRYAN: Well, hardly all my life—ever since I was about a year old.

DARROW: And good language, too. And you never took any pains to find out anything about the origin of languages?

BRYAN: I never studied it as a science.

DARROW: Have you ever, by any chance, read Max Mueller?

BRYAN: No.

DARROW: The great German philologist.

BRYAN: No.

DARROW: Or any book on that subject?

BRYAN: I don't remember to have read a book on that subject, especially, but I have read abstracts, of course, and articles on philology.

DARROW: Mr. Bryan, could you tell me how old the earth is?

BRYAN: No, sir, I couldn't.

DARROW: Could you come anywhere near it?

BRYAN: I wouldn't attempt to. I could possibly come as near as scientists do, but I had rather be more accurate before I give a guess.

DARROW: You don't think much of scientists, do you?

BRYAN: Yes, I do, sir.

DARROW: Is there any scientist in the world you think much of?

BRYAN: Yes.

DARROW: Who?

BRYAN: Yes, the bulk of them.

DARROW: I don't want that kind of an answer, Mr. Bryan. Who are they?

BRYAN: I will give you George M. Price, for instance.

DARROW: Who is he?

BRYAN: Professor of geology in a college.

DARROW: Where?

BRYAN: He was out near Lincoln, Nebraska.

DARROW: How close to Lincoln, Nebraska?

BRYAN: About 3 or 4 miles. He is now in a college out in California.

DARROW: Where is the college?

BRYAN: At Lodi.

DARROW: That is a small college?

BRYAN: I didn't know you had to judge a man by the size of his college; I thought you judged him by the size of the man.

DARROW: I thought the size of the college made some difference.

BRYAN: It might raise a presumption in the minds of some, but I think I would rather find out what he believed.

DARROW: You would rather find out whether his belief coincided with your views or prejudices or whatever they are, before you said how good he was?

BRYAN: I don't think I am any more prejudiced for the Bible than you are against it.

DARROW: Well, I don't know.

BRYAN: Well, I don't know either. It is my guess.

DARROW: You mentioned Price because he is the only human being in the world so far as you know that signs his name as a geologist, that believes like you do.

BRYAN: No, there is a man named Wright who taught at Oberlin.

DARROW: I will get to Mr. Wright in a moment. I am asking you about Mr. Price. Who publishes his book?

BRYAN: I can't tell you. I can get you the book.

DARROW: Don't you know? Don't you know it is Revell and Company in Chicago?

BRYAN: I couldn't say.

DARROW: He publishes yours, doesn't he?

BRYAN: Yes, sir.

STEWART: Will you let me make an exception? I don't think it is pertinent about who publishes a book.

DARROW: He has quoted a man that every scientist in this country knows is a mountebank and a pretender, and not a geologist at all.

JUDGE RAULSTON: You can ask him about the man, but don't ask him about who publishes the book.

DARROW: Do you know anything about the college he is in?

BRYAN: No, I can't tell you.

DARROW: Do you know how old his book is?

BRYAN: No, sir; it is a recent book.

DARROW: Do you know anything about his training?

BRYAN: No, I can't say on that.

DARROW: Do you know of any geologist on the face of the earth who ever recognized him?

BRYAN: I couldn't say.

DARROW: But you think he is all right? How old does he say the earth is?

BRYAN: I am not sure that I would insist on some particular geologist that you picked out recognizing him before I could consider him worthy if he agreed with your views.

DARROW: You would consider him worthy if he agreed with your views.

BRYAN: Well, I think his argument is very good.

DARROW: How old does Mr. Price say the earth is?

BRYAN: I haven't examined the book in order to answer questions on it.

DARROW: Then you don't know anything about how old he says it is?

BRYAN: He speaks of the layers that are supposed to measure age, and points out that they are not uniform and not always the same, and that attempts to measure age by these layers where they are not in the order in which they are usually found, makes it difficult to tell the exact age.

DARROW: Does he say anything whatever about the age of the earth?

BRYAN: I wouldn't be able to testify.

DARROW: You didn't get anything about the age from him?

BRYAN: Well, I know he disputes what you say, and I say there is very good evidence to dispute it—what some others say about the age.

DARROW: Where did you get your information about the age of the earth?

BRYAN: I am not attempting to give you information about the age of the earth.

DARROW: Then you say there was Mr. Wright, of Oberlin?

BRYAN: That was rather I think on the age of man rather than upon the age of the earth.

DARROW: There are two Mr. Wrights, of Oberlin?

BRYAN: I couldn't say.

DARROW: Both of them are geologists. Do you know how long Mr. Wright says man has been on the earth?

BRYAN: Well, he gives the estimates of different people.

DARROW: Does he give any opinion of his own?

BRYAN: I think he does.

DARROW: What is it?

BRYAN: I am not sure.

DARROW: What is it?

BRYAN: It was based upon the last glacial age, that man has appeared since the last glacial age.

DARROW: Did he say there was no man on earth before the last glacial age?

BRYAN: I think he disputes the finding of any proof, where the proof is authentic, but I had rather read him than quote him. I don't like to run the risk of quoting from memory.

DARROW: You couldn't say then how long Mr. Wright places it?

BRYAN: I don't attempt to tell you.

DARROW: When was the last glacial age?

BRYAN: I wouldn't attempt to tell you that.

DARROW: Have you any idea?

BRYAN: I wouldn't want to fix it without looking at some of the figures.

DARROW: That was since the Tower of Babel, wasn't it?

BRYAN: Well, I wouldn't want to fix it. I think it was before the time given in here, and that was only given as the possible appearance of man and not the actual.

DARROW: Have you any idea how far back the last glacial age was?

BRYAN: No, sir.

DARROW: Do you know whether it was more than 6,000 years ago?

BRYAN: I think it was more than 6,000 years ago.

DARROW: Have you any idea how old the earth is?

BRYAN: No.

DARROW: The book you have introduced in evidence fails you, doesn't it? [referring to the Bible]

BRYAN: I don't think it does, Mr. Darrow.

DARROW: Let's see whether it does. Is this the one?

BRYAN: That is the one, I think.

DARROW: It says BC 4004.

BRYAN: That is Bishop Ussher's calculation.

DARROW: That is printed in the Bible you introduced?

BRYAN: Yes, sir.

DARROW: And numerous other Bibles?

BRYAN: Yes, sir.

DARROW: Printed in the Bible in general use in Tennessee?

BRYAN: I couldn't say.

DARROW: And Scofield's Bible?*

BRYAN: I couldn't say about that.

DARROW: You have seen it somewhere else?

BRYAN: I think that is the chronology actually used.

DARROW: Does the Bible you have introduced for the jury's consideration say that?

BRYAN: Well, you'll have to ask those who introduced that.

DARROW: You haven't practiced law for a long time, so I will ask you if that is the King James version that was

---

* A version of the King James Bible, first published in 1909, that included chronological annotations by Cyrus I. Scofield.

introduced. That is your marking, and I assume it is.

BRYAN: I think that is the same one.

DARROW: There is no doubt about it, is there, gentlemen?

STEWART: That is the same one.

DARROW: Would you say the earth was only 4,000 years old?

BRYAN: Oh no, I think it is much older than that.

DARROW: How much?

BRYAN: I couldn't say.

DARROW: Do you say whether the Bible itself says it is older than that?

BRYAN: I don't think the Bible says itself whether it is older or not.

DARROW: Do you think the earth was made in six days?

BRYAN: Not six days of twenty-four hours.

DARROW: Doesn't it say so?

BRYAN: No, sir.

STEWART: I want to interpose another objection. What is the purpose of this examination?

BRYAN: The purpose is to cast ridicule on everybody who believes in the Bible, and I am perfectly willing that the world shall know that these gentlemen have no other purpose than ridiculing every Christian who believes in the Bible.

DARROW: We have the purpose of preventing bigots and ignoramuses from controlling the education of the

United States, and you know it, and that is all.

BRYAN: I am glad to bring out that statement. I want the world to know that this evidence is not just for the view. Mr. Darrow and his associates have filed affidavits here stating, the purpose of which, as I understand it, is to show that the Bible story is not true.

MALONE (FOR THE DEFENSE): Mr. Bryan seems anxious to get some evidence into the record that would tend to show that those affidavits are not true.

BRYAN: I am not trying to get anything into the record. I am simply trying to protect the Word of God against the greatest atheist or agnostic in the United States. [Prolonged applause.] I want the papers to know I am not afraid to get on the stand in front of him and let him do his worst. I want the world to know that agnosticism is trying to force agnosticism on our colleges and on our schools, and the people of Tennessee will not permit that to be done. [Prolonged applause.]

DARROW: I wish I could get a picture of those claquers.

STEWART: I am not afraid of Mr. Bryan being perfectly able to take care of himself, but this examination cannot be a legal examination, and it cannot be worth a thing, Your Honor. I respectfully except to it, and call upon Your Honor in the name of all that is legal to stop this examination, and stop it here.

HAYS (FOR THE DEFENSE): I rather sympathize with the General [Stewart], but Mr. Bryan is produced as a witness because he is a student of the Bible, and he presumably understands what the Bible means. He is one of the foremost students in the United States, and we hope to show Mr. Bryan, who is a student of the Bible, what the Bible really means in connection with evolution. Mr. Bryan has already stated that the world is not merely 6,000 years old, and that is very helpful to us. And where your evidence is coming from, this Bible, which goes to the jury, is that the world started in 4004 BC.

BRYAN: You think the Bible says that?

HAYS: The one you have taken in evidence says that.

BRYAN: I don't concede that it does.

HAYS: You know that that chronology is made up by adding together all of the ages of the people in the Bible, counting their ages. And now then, let us show the next stage from a Bible student, that these things are not to be taken literally, but that each man is entitled to his own interpretation.

STEWART: The court makes the interpretation.

HAYS: But the court is entitled to information on what is the interpretation of an expert Bible student.

STEWART: This is resulting in a harangue and nothing else.

DARROW: I didn't do any of the haranguing; Mr. Bryan

has been doing that.

STEWART: You know absolutely you have done it.

DARROW: Oh, all right.

MALONE: Mr. Bryan doesn't need any support.

STEWART: Certainly he doesn't need any support, but I am doing what I conceive my duty to be, and I don't need any advice, if you please, sir. [Applause.]

JUDGE RAULSTON: That would be irrelevant testimony if it was going to the jury. Of course, it is excluded from the jury on the point it is not competent testimony, on the same ground as the affidaviting.

HICKS: Your Honor, let me say a word right there. It is in the discretion of the court how long you will allow them to question witnesses for the purpose of taking testimony to the Supreme Court. Now we, as taxpayers of this county, feel that this has gone beyond reason.

JUDGE RAULSTON: Well, now, that taxpayers doesn't appeal to me so much, when it is only 15 or 20 minutes time.

DARROW: I would have been through in a half-hour if Mr. Bryan had answered my questions.

STEWART: They want to put in affidavits as to what other witnesses would swear, why not let them put in affidavits as to what Mr. Bryan would swear.

BRYAN: God forbid!

STEWART: It is not worth anything to them, if Your Honor please, even for the record in the Supreme Court.

HAYS: Is it not worth anything to us if Mr. Bryan will accept the story of creation in detail, and if Mr. Bryan, as a Bible student, states you cannot take the Bible necessarily as literally true?

STEWART: The Bible speaks for itself.

HAYS: You mean to say the Bible itself tells whether these are parables? Does it?

STEWART: We have left all annals of procedure behind. This is a harangue between Col. Darrow and his witness. He makes so many statements that he is forced to defend himself.

DARROW: I do not do that.

STEWART: I except to that is not pertinent to this lawsuit.

JUDGE RAULSTON: Of course it is not pertinent, or it would be before the jury.

STEWART: It is not worth anything before a jury.

JUDGE RAULSTON: Are you about through, Mr. Darrow?

DARROW: I want to ask a few more questions about the creation.

JUDGE RAULSTON: I know. We are going to adjourn when Mr. Bryan comes off the stand for the day. Be very brief, Mr. Darrow. Of course—I believe I will make myself clearer. Of course, it is incompetent testimony before the jury. The only reason I am allowing this to go in at all is that they may have it in the appellate courts, as showing what the affidavit would be.

BRYAN: The reason I am answering is not for the benefit of the Superior court. It is to keep these gentlemen from saying I was afraid to meet them and let them question me. And I want the Christian world to know that any atheist, agnostic, unbeliever, can question me any time as to my belief in God, and I will answer him.

DARROW: I want to take an exception to this conduct of this witness. He may be very popular down here in the hills. I do not need to have his explanation for his answer.

BRYAN: If I had not, I would not have answered the question.

HAYS: May I be heard? I do not want Your Honor to think we are asking questions of Mr. Bryan with the expectation that the higher court will not say that those questions are proper testimony. The reason I state that is this, your law speaks for the Bible. Your law does not say the literal interpretation of the Bible. If Mr. Bryan, who is a student of the Bible, will state that everything in the Bible need not be interpreted literally, that each man must judge for himself, if he will state that, of course, then Your Honor would charge the jury.

We are not bound by a literal interpretation of the Bible. If I have made my argument clear enough for the attorney general to understand, I will retire.

STEWART: I will admit you have frequently been difficult of comprehension, and I think you are as much to blame as I am.

HAYS: I know I am.

STEWART: I think this is not legal evidence for the record in the Appellate Courts. The King James version of the Bible, as Your Honor says...

JUDGE RAULSTON: I cannot say that.

STEWART: Your Honor has held the court takes judicial knowledge of the King James version of the Bible.

JUDGE RAULSTON: No sir, I did not do that.

STEWART: Your Honor charged the grand jury and read from that.

JUDGE RAULSTON: I happened to have the Bible in my hand, it happened to be a King James edition, but I will charge the jury, gentlemen, the Bible generally used in Tennessee, as the book ordinarily understood in Tennessee, as the Bible, I do not think it is proper for us to say to the jury what Bible.

STEWART: Of course, that is all we could ask of Your Honor. This investigation or interrogation of Mr. Bryan as a witness, Mr. Bryan is called to testify, was of the counsel for the prosecution in this case, and has been asked something, perhaps less than a thousand questions, of course not personal to this case, and it has resulted in an argument, and argument about every other question cannot be avoided. I submit, Your Honor, it is not worth anything in the record at all, if it is not legal testimony. Mr. Bryan is willing to testify and is able to defend himself. I accept it, if the court please, and ask Your Honor to stop it.

HAYS: May I ask a question? If your contention is correct that this law does not necessarily mean that the Bible is to be taken literally word for word, is this not competent evidence?

STEWART: Why could you not prove it by your scientists?

DARROW: We are calling one of the most foremost Bible students. You vouch for him.

MALONE: We are offering the best evidence.

McKENZIE: Do you think this evidence is competent before a jury?

DARROW: I think so.

JUDGE RAULSTON: It is not competent evidence for the jury.

McKENZIE: Nor is it competent in the Appellate Courts, and these gentlemen would no more file the testimony of Col. Bryan as a part of the record in this case than they would file a rattlesnake and handle it themselves.

DARROW, HAYS, MALONE: We will file it. We will file it. We will file every word of it.

BRYAN: Your Honor, they have not asked a question legally, and the only reason they have asked any question is for the purpose—as the question about Jonah was asked—for a chance to give this agnostic an opportunity to criticize a believer in the word of God; and I answered the question in order to shut his mouth, so that he cannot go out and tell his atheistic friends that I would not answer his questions. That is the only reason, no more reason in the world.

MALONE: Your Honor, on this very subject I would like to say that I would have asked Mr. Bryan—and I consider myself as good a Christian as he is—every question that Mr. Darrow has asked him, for the purpose of bringing out whether or not there is to be taken in this court only a literal interpretation of the Bible; or whether, obviously as these questions indicate, if a general and literal construction cannot be put upon the parts of the Bible which have been covered by Mr. Darrow's questions. I hope, for the last time, no further attempt will be made by counsel on the other side of the case, or Mr. Bryan, to say the defense is concerned at all with Mr. Darrow's particular religious views or lack of religious views. We are here as lawyers with the same right to our views. I have the same right to mine as a Christian as Mr. Bryan has to his, and we do not intend to have this case changed by Mr. Darrow's agnosticism or Mr. Bryan's brand of Christianity. [Prolonged applause.]

JUDGE RAULSTON: I will pass on each question as asked, if it is objected to.

DARROW: Mr. Bryan, do you believe that the first woman was Eve?

BRYAN: Yes.

DARROW: Do you believe that she was literally made out of Adam's rib?

BRYAN: I do.

DARROW: Did you ever discover where Cain got his wife?

BRYAN: No sir, I leave the agnostics to hunt for her.

DARROW: You have never found out?

BRYAN: I have never tried to find.

DARROW: You have never tried to find?

BRYAN: No.

DARROW: The Bible says he got one, doesn't it? Were there other people on earth at that time?

BRYAN: I cannot say.

DARROW: You cannot say? Did that never enter your consideration?

BRYAN: Never bothered me.

DARROW: There were no others recorded, but Cain got a wife. That is what the Bible says. Where she came from, you don't know. All right. Does the statement "The morning and the evening were the first day" and "The morning and the evening were the second day" mean anything to you?

BRYAN: I do not think it necessarily means a twenty-four hour day.

DARROW: You do not?

BRYAN: No.

DARROW: What do you consider it to be?

BRYAN: I have not attempted to explain it. If you will take the second chapter—let me have the book. The fourth verse of the second chapter says, "Those are the generation of the heavens and of the earth, when they were erected in the day the Lord God made the earth and the heavens." The word "day" there in the very next chapter is used to describe a period. I do not see that there is necessity for considering the words, "the evening and the morning" as meaning necessarily a twenty-four hour day in the day when the Lord made the heavens and the earth.

DARROW: Then when the Bible said, for instance, "And God called the firmament heaven, and the evening and the morning were the second day," that does not necessarily mean twenty-four hours?

BRYAN: I do not think it necessarily does.

DARROW: Do you think it does or does not?

BRYAN: I know a great many think so.

DARROW: What do you think?

BRYAN: I do not think it does.

DARROW: You think these were not literal days?

BRYAN: I do not think they were 24-hour days.

DARROW: What do you think about it?

BRYAN: That is my opinion—I do not know that my opinion is better on that subject than those who think it does.

DARROW: You do not think that?

BRYAN: No. But I think it would be just as easy for the kind of God we believe in to make the earth in six days as in six years or in six million years or in six hundred million years. I do not think it important whether we believe one or the other.

DARROW: Do you think those were literal days?

BRYAN: My impression is they were periods, but I would not attempt to argue as against anybody who wanted to believe in literal days.

DARROW: Have you any idea of the length of the periods?

BRYAN: No I don't.

DARROW: Do you think the sun was made on the fourth day?

BRYAN: Yes.

DARROW: And they had evening and morning without the sun?

BRYAN: I am simply saying it is a period.

DARROW: They had evening and morning for four periods without the sun, do you think?

BRYAN: I believe in creation as there told, and if I am not able to explain it, I will accept it.

DARROW: Then you can explain it to suit yourself. Mr. Bryan, what I want to know is, do you believe the sun was made on the fourth day?

BRYAN: I believe just as it says there.

DARROW: Do you believe the sun was made on the fourth day?

BRYAN: Read it.

DARROW: I am very sorry. You have read it so many times, you would know, but I will read it again.

"And God said, let there be lights in the firmament of the heaven, to divide the day from the night; and let them be for signs, and for seasons, and for days, and for years.

"And let them be for lights in the firmament of the heaven, to give light upon the earth; and it was so. "And God made two great lights; the greater light to rule the day, and the lesser light to rule the night; He made the stars also.

"And God set them in the firmament of the heaven,

to give light upon the earth, and to rule over the day and over the night, and to divide the light from the darkness; and God saw that it was good. And the evening and the morning were the fourth day."

Do you believe, whether it was a literal day or a period, the sun and moon were not made until the fourth day?

BRYAN: I believe they were made in the order in which they were given there and I think in dispute with Gladstone and Huxley on that point—

DARROW: Cannot you answer my question?

BRYAN: —I prefer to agree with Gladstone.

DARROW: I do not care about Gladstone.

BRYAN: Then prefer to agree with whoever you please.

DARROW: Cannot you answer my question?

BRYAN: I have answered it. I believe that was made on the fourth day, in the fourth day.

DARROW: And they had the evening and the morning before that time for three days or three periods. All right, that settles it. Now, if you call those periods, they might have been a very long time.

BRYAN: They might have been.

DARROW: The creation might have been going on for a very long time?

BRYAN: It might have continued for millions of years.

DARROW: Yes, all right. Do you believe in the story of the temptation of Eve by the serpent?

BRYAN: I do.

DARROW: Do you believe that after Eve ate the apple, or gave it to Adam, whichever way it was, that God cursed Eve, and at that time decreed that all womankind thenceforth and forever should suffer the pangs of childbirth in the reproduction of the earth?

BRYAN: I believe what it says, and I believe the fact as fully.

DARROW: That is what it says, doesn't it?

BRYAN: Yes.

DARROW: And for that reason, every woman born of woman, who has to carry on the race, the reason they have childbirth pains is because Eve tempted Adam in the Garden of Eden?

BRYAN: I will believe just what the Bible says. I ask to put that in the language of the Bible, for I prefer that to your language. Read the Bible, and I will answer.

DARROW: All right, I will do that: "And I will put enmity between thee and the woman." That referring to the serpent?

BRYAN: The serpent.

DARROW: "And between thy seed and her seed. It shall bruise thy head, and thou shalt bruise his heel. Unto the woman He said, I will greatly multiply thy sorrow and thy conception. In sorrow thou shalt bring forth children; and thy desire shall be to thy husband, and he shall rule over thee." That is right, is it?

BRYAN: I accept it as it is.

DARROW: Did that come about because Eve tempted Adam to eat the fruit?

BRYAN: I believe it is just as the Bible says.

DARROW: And you believe that is the reason that God made the serpent to go on his belly after he tempted Eve?

BRYAN: I believe the Bible as it is. And I do not permit you to put your language in the place of the language of the Almighty. You read that Bible and ask me questions and I will answer them. I will not answer your questions in your language.

DARROW: I will read it to you from the Bible: "And the Lord God said unto the serpent, Because thou hast done this, thou art cursed above all cattle, and above every beast of the field. Upon thy belly shalt thou go and dust shalt thou eat all the days of thy life." Do you think that is why the serpent is compelled to crawl upon its belly?

BRYAN: I believe that.

DARROW: Have you any idea how the snake went before that time?

BRYAN: No, sir.

DARROW: Do you know whether he walked on his tail or not?

BRYAN: No sir, I have no way to know. [Laughter.]

DARROW: Now, you refer to the cloud that was put in the heavens after the flood, the rainbow. Do you believe in that?

BRYAN: Read it.

DARROW: All right, Mr. Bryan, I will read it for you.

BRYAN: Your Honor, I think I can shorten this testimony. The only purpose Mr. Darrow has is to slur at the Bible, but I will answer his questions. I will answer it all at once, and I have no objection in the world. I want the world to know that this man, who does not believe in a God, is trying to use a court in Tennessee...

DARROW: I object to that.

BRYAN: ...to slur at it, and, while it requires time, I am willing to take it.

DARROW: I object to your statement. I am examining you on your fool ideas that no intelligent Christian on earth believes!

JUDGE RAULSTON: Court is adjourned until nine o'clock tomorrow morning.

H.L. MENCKEN is perhaps the foremost journalist in American history. Born in Baltimore, he wrote from the turn of the century until the late 1940's for *The Baltimore Sun*, and was known for his savage wit, erudite if salty language, and an iconoclastic outlook that saw through politicians and fads with fearless abandon. He was also the founder and editor-in-chief of the legendary news magazine *The American Mercury*, and the author of nineteen books. He died in Baltimore in 1956.

ART WINSLOW is a writer and literary journalist whose criticism frequently appears in the *Los Angeles Times*, the *Chicago Tribune, Bookforum,* and other publications. He spent many years as literary editor and executive editor of *The Nation* magazine, and is a past president of the National Book Critics Circle.